How to survive a nuclear emergency

2nd Edition

By Dr Keith Pearce

Contents

1. Introduction..1
2. How to survive a nuclear accident - quick overview.................................1
3. A little bit of background science 3
4. How could it affect me?...4
   4.1. Do I live or work near a nuclear site? 4
   4.2. How could I receive a radiation dose in a nuclear emergency?...........5
   4.3. Reduction of dose with distance downwind 7
   4.4. How can I minimise my radiation dose in an emergency? 8
   4.5. How do the authorities decide on which countermeasures to recommend?
   10
5. How will I know what to do?...11
   5.1. Prior information and the media...11
   5.2. Go in, Stay in, Tune in. 12
   5.3. Being prepared...12
   5.4. First actions...14
   5.5. Listening to advice 15
      5.5.1.   Introduction 15
      5.5.2.   Operating Company 15
      5.5.3.   Regulators...16
      5.5.4.   Police 16
      5.5.5.   Public Health England (PHE) Centre for Radiological, Chemical and
              Environmental Hazards (CRCE)..17
      5.5.6.   Food Standards Agency/Food Standards Scotland...............17
      5.5.7.   Local Authority...18
      5.5.8.   NHS...18
      5.5.9.   GP Services..18
      5.5.10. Media 18
6. The Countermeasures..19
   6.1. Shelter..19
      6.1.1.   What does sheltering mean?...19
      6.1.2.   What can we do when in shelter? 21
      6.1.3.   What can we eat and drink when in shelter? 21
      6.1.4.   How long would we be expected to shelter? 22
      6.1.5.   What happens to my children at school?...............................22
      6.1.6.   What do I do if I think that I, or one of my children, are
              contaminated?...23

6.2. Taking stable iodine tablets..24

6.3. Evacuation...26

    6.3.1.   What happens at the Rest Centres? 27

    6.3.2.   What are Humanitarian Assistance Centres?...........................30

    6.3.3.   What are Radiation Monitoring Units (RMUs)?......................31

6.4. Mass decontamination 33

    6.4.1.   How might the evacuation be managed? 34

    6.4.2.   What happens to my pets?..34

    6.4.3.   What happens to farm animals? 35

    6.4.4.   Return from Evacuation..36

7.   Caring for yourselves and your family after the event 39

8.   Summary..41

Appendix A: Nuclear Reactors, Accidents and Emergency Plans 42

A.1   How nuclear reactors work...42

A.2   A brief history of nuclear accidents..43

The Windscale Fire 43

The accident at Three Mile Island 44

Chernobyl..44

Fukushima...47

International Nuclear Event Scale (INES)...48

A.3   Can it happen here? 50

A.4   Reference Accident sequences in the UK 50

A.5   Emergency plans...52

Appendix B: Radiation and Radiation Protection 55

B.1.   What is radiation and radiation dose?...55

B.2.   Harm caused by radiation...58

B.3.   A brief history of Radiation Protection.......................................59

B.4.   Radiation Protection today...60

B.5.   What dose limits apply to me in an emergency?..........................61

Appendix C - Other types of event 64

C.1   Dirty Bomb 64

C.2   Orphan source...66

C.3   Nuclear satellite re-entry 68

C.4   Nuclear bomb...69

C.5   Nuclear risks in context..73

Appendix D - Key Radionuclides 78

D.1.Iodine..78

D.2.Caesium...79

D.3. Americium 79

D.4. Noble Gases 80

D.5. Plutonium 80

Appendix E Further reading 82

E.1 How to survive a nuclear war 82

E.2 Introduction to radiation 82

E.3 Prior information 83

E.4 Being prepared for emergencies 83

E.5 Countermeasures 84

E.6 Decontamination 85

E.7 Nuclear accidents 85

E.8 Nuclear emergency plans 86

E.9 Radiological protection 87

E.10 Lexicon 89

Appendix F - Your Home Resilience Plan 94

F.1 Risk Assessment 94

F.2 Flood planning 95

F.3 Utilities 96

F.4 Drinking water interruption 97

F.5 Lost phone 98

F.6 Local radio station 98

F.7 Family safe place / meeting place 98

F.8 Family documents 99

F.9 Special needs 99

F.10 Home Emergency Kit 100

F.11 Car Emergency Kit 101

F.12 Risk Assessment forms and family emergency plan templates 103

1. Introduction

Imagine the situation, it is midday, you are at home and the radio is on in the background. You notice a change. The DJ interrupted the last track and is speaking in an excited tone. You listen. She is talking about police advice to people within 3 km of the nuclear power station at Oldness-on-Sea to go indoors and stay there listening to the radio or television. What does it mean? What should you do? Listening to the radio for a few minutes more you soon realise that the DJ has got very little information so you turn the television onto a 24 hour news programme.

The TV channel has a banner headline saying "nuclear accident at Oldness-on-Sea, residents told to shelter". The message running along the bottom of the screen is repeating "The police have issued a warning to people near Oldness-on-Sea nuclear power station to shelter after an off-site nuclear emergency has been declared". Two presenters are sitting at a desk and one of them is showing where Oldness-on-Sea nuclear power station is and showing library footage of the site. It is obvious that not a lot is known about the event.

So what do you do? Where can you get advice to help you understand the situation? What is the best thing to do for you and your family? The nuclear industry provides leaflets giving advice to those that live near nuclear licensed sites but, although very good, these tend to be brief. This book tries to give fuller answers to the questions about personal and family safety in a nuclear emergency in an accessible style.

References are given throughout the book in the form of links to internet pages and further reading is recommended in Appendix E.

Radiological protection and emergency planning are all areas plagued by jargon so a list and brief explanation of some of the key terms is provided in a Lexicon at the end of the book.

This book is mainly about understanding and surviving a nuclear accident at a nuclear licensed site in the UK but there is an appendix on other types of radiation events (Appendix C) and in Appendix F there is a Section that guides you towards writing a household contingency plan.

2. How to survive a nuclear accident - quick overview

To survive a nuclear accident you will want to achieve a number of things:
- To be sure that your family and friends are safe;

- Be sure that your health and the health of your family and friends has not been affected;
- To be allowed to get on with your life with as little disruption as possible.

The first of these is really about understanding the situation and about communication.

A very brief coverage of the background science is given in Section 3 with more details being given in Appendices A and B.

Section 4 of this book outlines how you could be affected by a nuclear emergency if you are downwind of a site during an accident. It briefly describes how you might be able to reduce your radiation dose by taking countermeasures such as staying in shelter, evacuating and/or taking stable iodine and how the authorities decide what countermeasures to advise.

Do you know how to contact your family if there is an emergency? You may want to be reunited with your family as quickly as possible. Problems may arise where some members of a family are in the affected zone and some are not. If you have children at school you can rest assured that the school will look after them until you can be reunited even if that is later than the usual school closing time. A Family Emergency Plan, made and agreed before the event, can help a family cope with a range of crises. See *Being Prepared* in Section 5 and Appendix F which outlines a home contingency plan..

If you are in the affected area you can help reassure your friends and family by texting them, phoning them or by posts on any Social Media (Facebook, Twitter etc.) accounts that people associate you with. You are asked to use phones in the affected area as little as possible so text is better or you could phone one friend or family member outside the area and ask them to contact and reassure the others.

There are a number of things that the authorities can recommend you do that will help to reduce your exposure to radiation. For short releases with little or no warning you are best sheltering in your home and allowing the radioactive gas and dust to blow over the house and away from you. For more severe, or for longer lasting releases, you may be best to evacuate your home and go to somewhere more distant from, or upwind from, the release. If the event involves an operating nuclear reactor then the taking of stable iodine, usually in the form of tablets, can reduce the dose to thyroid and reduce health effects, particularly in children and young adults. These countermeasures, and how the Local Authority will support you at the time, are explained in Section 6. After the release has stopped attention turns to the contamination in the environment and in locally grown food.

History has shown that the effect of a nuclear accident on health is complex. Unless it is a very severe accident and people are close to the scene it is unlikely that anyone will suffer noticeable health effects directly from radiation. People receiving lower doses of radiation will have a slightly increased change of getting cancer later in their lives but the increase is so slight compared to the normal rate of cancer that it would be very difficult to prove any radiation effects in society at all. See Appendix B, Radiation and Radiation Protection.

However, Chernobyl and Fukushima have shown that some people suffer Post Traumatic Stress Disorder[1] and other stress related problems which can have a very real damaging effect on their health. This is particularly true of populations who are displaced from their homes either temporarily or permanently by the event or stay in the area but live in fear of the radioactive material contaminating the environment. Some information on staying healthy after the event are given in Section 8.

On the day of the accident you need to understand what is happening and to understand that you can take certain actions to reduce your radiation dose and the dose to your family. After the event you need to be reassured that you and your family are safe. That is a matter of understanding any additional radioactivity in the area you live and work, understanding the very low likelihood of it doing you any harm and learning to relax and be comfortable with the situation. That may be easier said than done.

3. A little bit of background science

Nuclear reactors work by splitting atoms to release energy which is turned into electricity. An unwelcome side product of this process are radioactive materials. These have within them unstable atoms that "decay" giving out "ionising radiation" which can affect the cells in your body and, if there is enough radiation, can make you sick or they can start a cancer forming in your body.

The designers and operators of nuclear power stations are very careful to keep the radioactive material trapped away from the environment so the chances of a significant release is very small indeed. There is more detail about nuclear reactors in Appendix A.1 and previous accidents in Appendix A.2. A discussion about UK nuclear safety is given in Appendix A.3 and the reference accidents used for planning purposes in the UK outlined in Appendix A.4.

---

[1] http://www.nhs.uk/conditions/post-traumatic-stress-disorder/Pages/Introduction.aspx

But we all know that accidents can happen so there are emergency plans at all nuclear sites that allow the operators to respond quickly to any event, to prevent or minimise a release and to warn those living in the area to take precautions as necessary. There is more detail about emergency plans in Appendix A.5.

The potential effects of radiation on the body are briefly summarised in Appendix B.

4. How could it affect me?

4.1. Do I live or work near a nuclear site?

A listing of the nuclear sites subject to the emergency planning requirements of REPPIR and their associated Detailed Emergency Planning Zones (DEPZs) can be found on the ONR website[2]. If you live in one of these zones the regulators believe that you could be affected by a radiation emergency (Appendix A). Figure 1 - Regulated sites (From ONR) shows a map of regulated sites in the UK.

---

[2] http://www.onr.org.uk/depz.htm

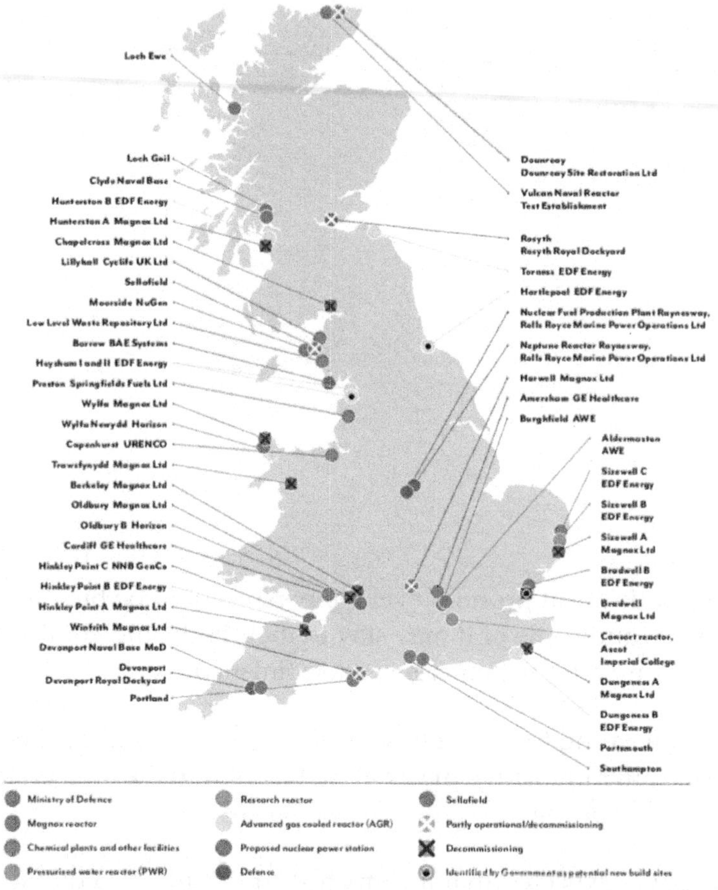

Figure 1 - Regulated sites (From ONR)

4.2. How could I receive a radiation dose in a nuclear emergency?

The immediate effect of an off-site nuclear emergency is likely to be a cloud of radioactive material drifting downwind away from the site and spreading out as it goes or, at least, the threat of such a cloud. This radioactive material gives off ionising radiations which, if they hit your body, can cause damage to body tissues which can then lead to health effects.

Anyone near enough to this cloud of material would be subject to a radiation dose from the cloud with the dose rate depending on what was in the cloud, how much of it was there, how far it was from the person and how long it was there for.

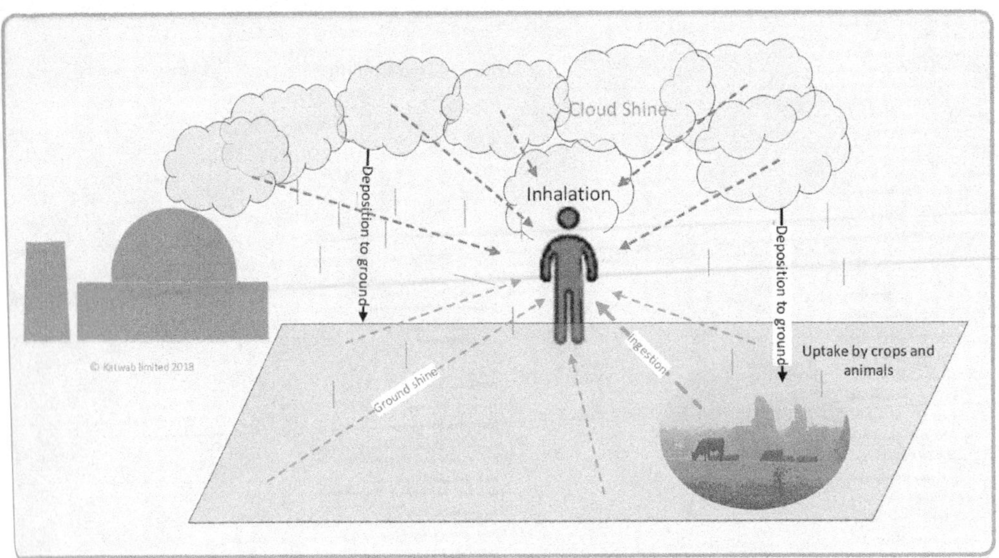

Figure 2 Dose Routes

More importantly, if the cloud was at ground level the person could be breathing in the radioactive dust or gas and some of it may stay in their body and continue to give them a radiation dose long after the emergency has finished.

Later dose routes result from eating food that has become contaminated by the plume and radiation dose from radioactive material that landed in the area on the ground, plants and trees and buildings.

In most accidental releases it is expected that the greatest contribution to public dose will arise from inhalation of the plume. Inhalation dose can be reduced by sheltering (when there are lower airborne levels inside a building than outside), by evacuation and, in some cases, taking stable iodine. Face masks, other than those designed for the purpose and well fitted, are of debatable value[3]. N95 face masks are available on the market. These are supposed to be able to block 95% of particles of 0.3 micron or larger but only achieve this if well fitted - they don't work nearly so well if you have a beard.

The three ways to reduce the shine dose (direct shine, ground shine and cloud shine) are time, distance and shielding.

Time. You continue to build up a radiation dose all the time you are within the range of a source so it is best to reduce this time - move away from the source of radiation as quickly as possible.

---

[3] https://blogs.cdc.gov/niosh-science-blog/2018/01/04/respirators-public-use/

Distance. The radiation dose from a source reduces with distance. If you move further away from the source the dose rate is reduced just as the heat from a fire seems less from further away.

Shielding. If you put material between yourself and the source then some of the radiation will be absorbed by the material and consequently your dose will be reduced. Different radiations have different ranges in material with alpha and beta radiations being stopped by thin layers but gamma radiation requiring a very thick shield to remove all of the radiation.

Decontamination. Dealing with contamination which is loose radioactive material in the form of a dust or liquid, is more complex. If you get it on your skin or clothes it can continue to emit radiation and give you a radiation dose that you cannot walk away from. To deal with contamination properly you have to remove it, a process called *decontamination*. This is achieved by either by removing the clothes it is on, or by washing it off and disposing of it carefully (See section 6.1.6).

If you get contamination into your body either by breathing it in, by eating it (this can be done by eating contaminated food or by touching your lips with contaminated hands) or by being wounded by a contaminated object (often called contaminated puncture wounds by health physicists) then it remains in your body until it decays or is excreted. Radioactivity inside the body is worse than radioactivity outside the body which is why care is taken to reduce inhalation (filtered face masks are used in contaminated areas on nuclear power stations) and care is taken to avoid ingestion (eating and drinking are usually forbidden in contaminated areas).

4.3. Reduction of dose with distance downwind

It is important to realise that the dose you might be exposed to as a result of an atmospheric discharge from a site depends very strongly on where you are. Activity released to the air will tend to follow the wind, spreading out sideways and upwards as it goes, depositing on the ground and other surfaces and being subject to radioactive decay. This means that further away from the source the airborne concentration is less; significantly less.

NRPB-W19[4] provides methods and data for estimating dose to the public following accidental releases. From figure 13 in this document it is possible to estimate the change in dose with distance for a ground level release lasting half an hour for common weather conditions (Table 1).

---

[4]

http://webarchive.nationalarchives.gov.uk/20100303233541/http://www.hpa.org.uk/webw/HPA web&HPAwebStandard/HPAweb_C/1248854051111?p=1158945066506

Distance (downwind plume centre) /m	Dilution Factor $Bq.m^{-3}/Bq.s^{-1}$	Ratio to 100m	1/ratio
100	9.0×10^{-4}	1.000	1
200	3.0×10^{-4}	0.3333	3
500	5.0×10^{-5}	0.0556	18
1000	1.8×10^{-5}	0.0200	50
10,000	4.0×10^{-7}	0.0004	2250

Table 1 Dilution factor with distance

Note: Figures taken from NRPB-W19 for a ground level release in Cat. D weather conditions.

The dilution factor gives the airborne concentration ($Bq.m^{-3}$) at the position in question for a release rate of 1 $Bq.s^{-1}$.

This shows that, under these circumstances, someone at 10 km would get about 2000 times less dose than someone at 100 m and someone at 500 m would get about 20 times less dose than someone at 100 m.

Exposure also drops off quickly as you move away from the plume centre.

It is those locations that are near the site and directly downwind that need priority for achieving countermeasures. Those further out can take lessor countermeasures, such as shelter instead of evacuation, or wait until the inner zones are safely evacuated before leaving their homes, if evacuation should be deemed necessary.

4.4. How can I minimise my radiation dose in an emergency?

It is likely that in a nuclear emergency the greatest doses to the public will result from the breathing in of the radioactive material and the second most important route will be the cloud dose from being in or near the cloud as it passes. The initial countermeasures to protect the public are aimed at reducing these.

The four main countermeasures are:

Evacuation

If you have warning that a big release is going to happen or know that the current release will last a long time the most effective countermeasure is to evacuate people in the plume's pathway. This removes them from any radiation dose but is very disruptive of their day.

Shelter

If people can stay inside a reasonably solid building such as a typical house they can reduce their inhalation dose. This is because the radioactive material is blown around their shelter rather than getting into it. They also reduce their cloud and ground shine doses. This is because they are further away from the airborne source and the building provides shielding.

Stable Iodine

Radioactive iodine is produced by nuclear reactors and may be released by a range of accidents. If inhaled it concentrates in the thyroid gland in the neck and can cause problems with this organ later. It has been shown that if you swallow a relatively large amount of non-radioactive iodine (stable iodine) then the thyroid gland becomes full of iodine and does not hold on to the radioactive iodine. Instead this quickly passes through the body causing much less harm. In some documents stable iodine is referred to as PITs (Potassium Iodate Tablets).

Radio-iodine is short lived so decays away within days of a reactor being shut-down. For this reason stable iodine tablets are only useful in the event of an accident at an operating reactor.

Food controls

Eating food that is contaminated with radioactive material leads to an internal radiation dose. In the event of a nuclear accident the government will recommend that people think about where their food has come from and don't eat food that could be contaminated. Generally this means food grown in fields, allotments and gardens around the affected site and food sold in open markets where the plume may have deposited activity. Tinned and packaged foods are generally fine. Extra rinsing of vegetables may be a good idea. Radioactivity which is eaten is more dangerous than radioactivity outside the body[5] so contamination on food is more worrying than contamination on the ground and buildings. Because of this food controls reach further from the affected site than advice to shelter. The Food Standards Agency (Food Standards Scotland in Scotland) may ban the sale of food produced in affected areas.

---

[5] There are a number of reasons for this (a) because it is in the body you cannot walk away from it or be shielded from it. Some types of radioactivity will lodge in the body for a very long time and carry on giving you an additional radiation dose, others pass through quickly. (b) The short range radiations (alpha and beta) which are largely harmless outside the body now hit sensitive living material and cause damage. (c) A larger fraction of the energy is deposited in the body rather than elsewhere.

4.5. How do the authorities decide on which countermeasures to recommend?

In the acute phase of a nuclear accident, that is, when the radioactive plume is being released, the three main countermeasures are shelter, stable iodine and evacuation. Food controls, and decontamination come later. The authorities have to decide which ones, if any, are the best for you under the circumstances on the day.

Estimates have been made about the fraction of dose that each countermeasure can prevent. In the scientific literature this is referred to as "averted dose" or saved dose. Thought has also been given to how much disruption would be caused by each countermeasure and the savings compared to the costs. The National Radiation Protection Board (NRPB), now part of Public Health England, has considered this cost/benefit and produced guidance called Emergency Reference Levels (ERLs)[6] which suggest levels of dose saving that must be achievable for a countermeasure to be worthwhile.

On the day of the accident people will be trying to estimate the dose those downwind of the event might receive. They will compare these estimates to the ERLs to see which, if any, of the three countermeasures or combinations of them would be better for members of the public.

Many of the emergency plans in the UK will automatically recommend shelter and the taking of stable iodine for those closest to the site on declaration of an off-site nuclear emergency. This ensures that the countermeasures are in place as quickly as possible so that they have the most effect. In many cases it is expected that the countermeasure can be in place before the release starts because the accident sequences give some warning time. The distances out to which automatic countermeasures are taken are agreed with the regulator and are based on potential accidents at the upper end of what is reasonably foreseeable.

The police coordinate the protection of the public taking advice from a wide range of organisations. See emergency plans later in this document (Section A.5).

---

[6] https://www.gov.uk/government/uploads/system/uploads/attachment_data/file/423879/NRPB_Vol_1_No_4_1990_-_for_website.pdf

5. How will I know what to do?

5.1. Prior information and the media

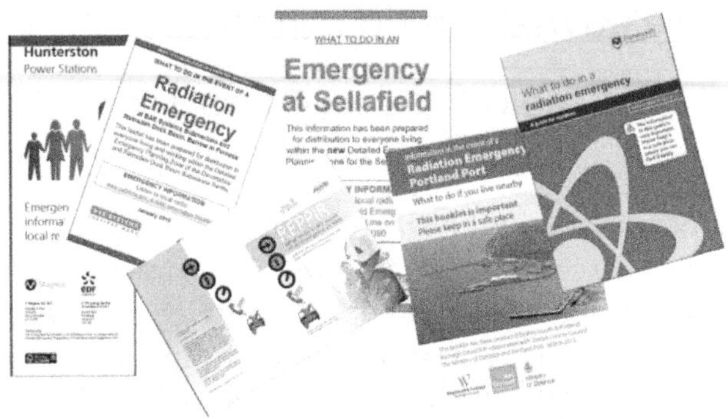

If you live within the area that the regulators think is close enough to the site to be
affected by an off-site nuclear emergency (i.e. within the Detailed Emergency
Planning Zone) then the operators will have tried to send you information about
the local emergency plan. This is a requirement of the Radiation (Emergency
Preparedness and Public Information) Regulations 2001 Regulation 16. There are
several versions of this available on the internet (for example BAE Systems Barrow
in Furness[7], AWE[8], Sellafield[9], Hunterston[10], Portsmouth[11], Portland[12], and
Dungeness[13]).

The information provided includes basic information about radioactivity and its
effects on the body; the types of radiation emergency and their potential
consequences, the steps that will be taken to alert, protect and assist you,
information on the activities you should take and the authorities with
responsibility within the plan.

---

[7] http://www.cumbria.gov.uk/eLibrary/Content/Internet/536/5858/42053142012.pdf

[8] http://www.awe.co.uk/app/uploads/2014/07/REPPIR_May-2013.pdf

[9] http://www.sellafieldsites.com/wp-content/uploads/2012/07/What-to-do-in-an-Emergency-Booklet-Phase-1-Shelter-FINAL-061015.pdf

[10] https://www.edfenergy.com/sites/default/files/2013-hunterston.pdf

[11]https://www.portsmouth.gov.uk/ext/documents-external/cou-pmouthwhattodoradiationemergency.pdf

[12] https://www.dorsetforyou.gov.uk/media/163294/Information-in-the-event-of-a-Radiation-Emergency-in-Portland-Port/pdf/Radiation_booklet_04_02_2013_web_vers.pdf

[13] http://www.edfenergy.com/sites/default/files/2013-dungeness.pdf

5.2. Go in, Stay in, Tune in.

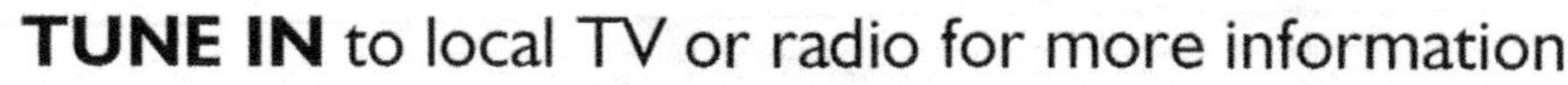

The usual advice for coping with emergencies in the UK is summarised by the phase "go in, stay in, tune in". This advice comes from the UK Government and is given in a leaflet[14] which was widely distributed. In essence this advice gets you into a building where you benefit from the shielding it offers you from harm (which it does if the agent of harm is wind borne such as smoke from a fire, chemicals from a leak or radioactivity from a nuclear release or if it is some other forms of risk such as a roaming gunman).

The "stay in" advice is to keep you in that safe place and avoid opening doors too often as, of course, opening doors gives the contaminant in the air outside a chance to get indoors.

"Tune in" is there because BBC local radio will be broadcasting news about the event and guidance from the authorities about how you should behave in order to minimise your harm. This might be to stay indoors until the event is over and then ventilate your house and take other precautions to reduce harm.

The prior information you receive from the site operator will tell you the details of your local radio stations that will be transmitting advice.

Some nuclear site operators have a system that allows them to send a message to any phone in the area that is registered. If you live in the area covered by one of these systems it is a very good idea to register. Details on how to do this will be sent to you by the operator periodically.

5.3. Being prepared

Accidents and emergencies can happen at any time and to anyone. You are more likely to come out of an emergency well if you are prepared. There is a lot of advice on the steps individuals and families can take to prepare for emergencies.

These generally consist of:

---

[14] http://www.cheshirefire.gov.uk/Assets/preparingforemergencies.pdf

Being prepared to stay in the home for a period of time, possibly without one or more utilities (gas, electricity and/or water).

This involves having a stock of food and water which would allow you to survive for a few days without going out; battery, or better still manually, powered torches to be able to move around safely after dark; a battery or manually powered radio to be able to listen to advice from the authorities; a first aid kit and supplies of any essential medications.

Having an old fashioned phone that plugs directly into the phone socket in the wall is also a good idea as cordless ones will not work if there is a cut in the mains electricity. Mobile phones are obviously very useful while the batteries last so you may wish to use them wisely but sparingly if the mains power is cut to make the batteries last longer.

Being prepared to leave the home at short notice. This involves knowing where the things you might want to take with you are stored so that you can pack quickly. Recommended items include clothing and food, taking into account the time of year and the weather and temperatures expected; cash and credit cards; medications and a first aid kit; spare pair of glasses; sleeping bags or blankets in case an overnight stay is required; contact details of friends and family; mobile phones and chargers; toilet paper; plastic bags including bin bags; torches and batteries; games and books particularly for children as there could be long periods of boredom ahead. Some of these can be pre-packed in an "emergency kit" ready to go when needed.

Having a family emergency plan in which you gather together each other's contact details and agree where you might meet if the family home is unobtainable; know how to make the house safe if you are leaving it for a few days at short notice; know where your essential documents such as birth certificates, passports and insurance details are kept; know how you are going to manage pets in such a situation. Appendix F shows how to prepare a home contingency plan.

Having an "emergency friend" who is someone you can trust who lives at least a few miles from you who will help you and your family should the need arise maybe by allowing you to stay with them for a short while or by keeping copies of your important documents and house keys. This may be where you agree to meet if need be.

Being ready to look after neighbours. Are any of your neighbours less able to care for themselves? Do you know their telephone number? Do you know how to contact their family if need be?

For those living or working in a nuclear site's DEPZ it would be helpful to know where the information provided by the site is be found and where your stable iodine tablets are stored if you have been given them (operating reactor sites only).

There is a lot of good advice on the internet about preparing families to cope with crises. For example: UK Government[15], Gloucestershire[16], Safer Scotland[17], British Red Cross[18] and Scottish Highlands[19]. An American version[20] and a Canadian Version[21] are interesting.

5.4. First actions

If you hear about a nuclear emergency near to you the first thing to do is to listen carefully to the information on your local radio station. Has a release of radioactive material started or are they worried that one might start soon? Are you close to and downwind of the affected site? What are they advising?

If a release has started and you are downwind of the site you should go indoors and close any doors and windows. Switch off any heating or air-conditioning. This is to keep as much of the radioactive dust and gases out of your house as possible. While you are in your home you can carry on as usual but keep listening to the local BBC radio channel for information.

Are they advising people in your area to take stable iodine tablets (sometimes called PITs) now? If they are, and you have a stock in the house then follow the instructions in the pack and provide the right dose to everyone in your house. If you live in an area where the tablets are distributed on the day then listen to the radio to find out how and when yours will arrive or where you have to go to collect them.

If there is a warning period you may be advised to find the tablets and read the instructions but not take them yet.

If the authorities are advising you to evacuate you should listen carefully to the advice given. What routes out are they recommending? Do you go in your own car if you have one or is transport being provided? Think carefully about what to take.

---

[15] https://www.gov.uk/government/publications/preparing-for-emergencies/preparing-for-emergencies

[16] http://glosprepared.co.uk/be-prepared/preparing-your-family/

[17] http://www.readyscotland.org/at-home/

[18] http://www.redcross.org.uk/What-we-do/Preparing-for-disasters/How-to-prepare-for-emergencies

[19]

http://www.highland.gov.uk/downloads/file/6342/preparing_for_emergencies_in_the_highlands_and_islands

[20] https://www.ready.gov/prepare-for-emergencies

[21] http://nuclearsafety.gc.ca/eng/resources/emergency-management-and-safety/protecting-your-family-during-a-nuclear-emergency.cfm

5.5. Listening to advice

5.5.1. Introduction

In the event of a nuclear emergency many people and organisations may be giving advice in the media and social media. Be selective in which ones you believe and act upon. When hearing advice about how to behave during and after a nuclear accident you need to consider the source of advice:

- Is the person/organisation knowledgeable in the fields of radiation protection, nuclear accident progression and the site's emergency plan?

- Does the person/organisation have full access to current information about the event and about the dispersion of radioactivity?

- Have they had the opportunity to discuss it with other experts who have different angles on the situation?

Only when all three of these are true can the information be considered to be fully trustworthy (and even then, of course, the advice could be overtaken by changing circumstances). Other sources may be well-meaning, attention seeking or working to another agenda but will tend to be less well informed and should be reacted to with care.

You might consider listening to several supposedly reliable sources of information and comparing them (triangulation). If they are sufficiently similar you may start to trust the content and reliability.

The following section looks at who might be trusted to understand the situation and report sensible advice.

5.5.2. Operating Company

The site operating company have the responsibility to raise the alarm as quickly as possible after discovering an event which has led to, or may lead to, a significant release of radioactivity. There are planned call out chains and systems to alert all those who need to know at all sites and these are tested regularly.

The operating company is then responsible for managing the on-site situation and giving information to the off-site authorities about what is happening and how they intend to bring the situation under control. Within the information they give are running reports on the estimated rate at which radioactive material is being released and estimates of how the release rate will vary in the future.

Many plans have automatic countermeasures and it may be the operator's responsibility to activate these and send the initial alert to the affected public. For some sites it is the local police who promulgate the initial alert to the public.

The operating company will usually send a management and technical team to the Strategic Co-ordinating Centre (SCC) to provide information, interpretation and advice to the authorities there.

If you live within a countermeasure zone you will have had prior information on how to respond. If you receive an alert in the manner you had been warned about then be ready to follow that advice. Certainly switch your radio or television to local news to seek confirmation of the event.

5.5.3. Regulators

The ONR role in a major civilian event is summarized as to *"monitor and record the operator's actions; take enforcement action if appropriate; and provide advice to relevant authorities"*[22].

ONR will have representatives in many of the response centres and may have personnel on the affected site. These representatives will report the situation as they see it to the ONR control room (RCIS) for interpretation and action.

An ONR representative may talk to the media and be interviewed on TV and Radio. The ONR have a thorough understanding of the nuclear industry and radiation protection and will be well informed of the current situation. An official ONR representative talking to the media will be highly trustworthy and will be reporting their best and knowledgeable estimate of the risks posed by the situation.

The environment agency will be represented in the system and will be monitoring the potential for environmental harm and providing advice on how it can be minimised. They may provide advice to the public via press releases and interview.

5.5.4. Police

The police chair the Strategic Coordination Centre. This is a role in which they have experience and for which they train and exercise. The particular chairman on a given day may not be particularly familiar with the dangers of radiation but they will have good advice to hand and will be able to debate the issues.

---

[22] http://www.onr.org.uk/operational/inspection/onr-ep-in-003.pdf

Again, a police spokesman talking to the media, will be giving an unbiased opinion based on discussions with a wide range of knowledgeable people. As chair of the co-ordination group, they will recommend the countermeasures that you should take.

5.5.5. Public Health England (PHE) Centre for Radiological, Chemical and Environmental Hazards (CRCE)

Public Health England's Centre for Radiation, Chemical and Environmental Hazards (CRCE) maintains emergency response arrangements and facilities to be ready for a wide range of radiological and nuclear emergencies that might have an effect on public health. These arrangements cover all of the UK and UK dependent territories.

CRCE would provide public health advice and information relating to the radiological protection aspects of the emergency to colleagues at PHE Centres and other responding organisations including central government.

In an emergency they have representation at many of the key command centres including the SCC and COBR and a crisis centre in Chilton to co-ordinate their efforts.

Advice from PHE-CRCE would be knowledgeable (they have a good understanding of radiation protection and the nuclear industry), informed (on the day they would have close information ties with the operators, local authorities and other responding organisations) and unbiased (their role, as part of PHE, is to *"protect and improve the nation's health and wellbeing, and reduce health inequalities"*).

5.5.6. Food Standards Agency/Food Standards Scotland

In an emergency the FSA would provide advice to the SCG/STAC and responders on food contamination issues based on their own measurements of radioactivity in food and the reports of dispersion of radioactivity. They would take action to ensure that contaminated food which may pose a risk to human health does not enter the food chain and they would provide public advice and information, on food safety.

Their advice on whether or not food and water were safe to consume will be based on their understanding of the situation although if in doubt, and alternative supplies are available, they may knowingly overstate the situation. Their advice not to consume certain foods and drinks should be followed.

5.5.7. Local Authority

The local authority will be represented in the Strategic Coordinating Group and will be providing many of the services to those directly and indirectly affected by the emergency. Their advice on how to shelter, when to evacuate, how to evacuate and where to go to will be framed in consultation with the police and should be followed.

People in shelter at home or in offices with particular needs (home help, medical care etc.) should contact the local authority help line, the national health help line, their medical centre/Doctors' surgery or the emergency number (999) depending on the situation.

5.5.8. NHS

The National Health Service help lines will be able to give information and advice to people concerned about the effects of a nuclear accident but it is likely that, at least initially, it will be very general advice rather than specific to the situation on the day.

5.5.9. GP Services

Your GP Services may not know a great deal about radiation and the effects that it may have on human health but their education and experience mean that if there is an emergency in their area they will quickly learn the basics and be able to provide advice and support in the days and weeks following any potential exposure.

If an area is widely contaminated following a nuclear accident it is very likely that the local medical services will be involved in debating any clean-up priorities and helping to determine the advice to be given to people who continue to live in the area.

5.5.10.   Media

History has shown that in complex emergencies it is very easy for different organisations to present the event in different ways and the media and public can quickly become concerned that they are being told half-truths or being deliberately misled. Where information does not seem to be forthcoming some elements of media and, particularly social media, can fill the gaps with rumour and conjecture; an example of where *falsehood flies, and truth comes limping after it*[23].

---

[23] Swift (1667–1745) Political lying. http://www.bartleby.com/209/633.html

In an attempt to provide information to the public and media as quickly, accurately and openly as possible a Media Briefing Centre (MBC) will be set up near to the Strategic Coordination Centre (SCC). Normally, the Police Commander chairing the Strategic Coordinating Group (SCG) will, through the Strategic Media Advisory Cell (SMAC) or Media Communications Cell (MCC) have overall responsibility for the co- ordinated management of the Media Briefing Centre, which forms the working/ briefing area for the media.

When you see the Chair of the SCG (a senior policeman), a senior member of the ONR and a senior person from PHE share a platform to update the media you can be sure that you are seeing up to date, knowledgeable and open briefings.

6. The Countermeasures

Countermeasures are things that you do to reduce your radiation dose. The main ones are shelter, stable iodine and evacuation.

6.1. Shelter

6.1.1. What does sheltering mean?

Government emergency planning advice[24] states that shelter in place should be considered for no-notice events when:
- there is no time to undertake an evacuation before the hazard arrives;
- going outside would expose people to greater harm or dangerous conditions; or
- the immediate risk is unclear.

Shelter will be recommended when the potential dose saving outweighs the inconvenience imposed and where it is not considered to be a good idea to evacuate. The reasons why shelter might be preferred to evacuation include:
- the potential dose saving is not sufficient to merit evacuation;
- the fear that people will be caught in traffic jams under the plume and thus increase their dose or be caught in traffic jams elsewhere and have a miserable time for no good reason.

Shelter has been shown to be better than evacuation in some circumstances. In a report on learning from experience[25] the NEA state that *"In 1973, a sudden failure occurred in an anhydrous ammonia storage tank at Potchefstroom in South Africa. Workers in a building 80 m from the release survived, but people who left their houses 180-200 m from it died"*. It is unlikely that the levels of ionising radiation off-site will

---

24
https://www.gov.uk/government/uploads/system/uploads/attachment_data/file/274615/Evacu ation_and_Shelter_Guidance_2014.pdf
[25] https://www.oecd-nea.org/rp/pubs/2018/7308-all-hazards-epr.pdf

---

reach fatal levels, as it did for this chemical, but it is wise to carefully consider official advice even if you are tempted to leave the area.

Shelter may be used with or without stable iodine depending on the situation. Stable iodine is only useful in the event of an operating reactor accident. The National Radiation Protection Board (NRPB) (now part of public Health England) states that[26] *"in this context sheltering refers to staying indoors, with doors and windows closed and ventilation systems turned off. It provides protection from external irradiation from radioactive material in the air and that deposited on the ground, and from inhalation of radioactive material. Typically dose reductions for a solidly built and reasonably airtight UK housing are by a factor of ten for external exposure and by a factor of three for the inhalation of particles"*. It claims little effectiveness for shelter in reducing the inhalation of vapours such as elemental iodine and warns that the effectiveness quoted is based on a release of up to a few hours.

The Dungeness B advice leaflet from EdF, which is fairly typical, suggests that people at home should:
- Go indoors;
- Bring domestic pets indoors;
- Leave all farm animals where they are;
- Close all outside doors and windows;
- Switch off any ventilation or extractor fans;
- Tune into your local radio/TV channel and listen for any further instructions.

In addition some operators ask you not to use the telephone unless absolutely necessary. This is because there are concerns that the telephone network will become congested and emergency calls not get through.

Some prior information leaflets contain the advice <u>not</u> to leave the area unless advised to by the emergency services. This is generally good advice at all sites since most buildings provide more protection than a car and there is a possibility that you will be caught in traffic jams under the plume of radioactive material. Any traffic jams caused by people trying to leave the area will make the response to the incident harder to manage.

While social media is a very good way of finding out if friends and family have been affected it is also the source of rumour and exaggeration. If your mobile phone is working then use social media wisely to gather news of friends and family and to access official advice. Always consider the source of any information carefully before deciding whether or not to trust it.

---

[26]

https://www.gov.uk/government/uploads/system/uploads/attachment_data/file/423879/NRPB _Vol_1_No_4_1990_-_for_website.pdf

6.1.2. What can we do when in shelter?

When you have got everyone indoors and the doors and windows are shut you can do pretty much anything that you would normally do. It is a good idea to listen to local radio for updates.

If the release had started before you came inside take off your outer layers of clothes and carefully bag them. This will remove most of any contamination you may have picked up.

If it looks like the event might continue for a while consider if you have enough food and water for everyone. Does everyone have any medicines that they might need in the next few hours? If medicine supplies might be a problem try to phone your doctors' surgery or the help line that the local radio station is advertising for help and advice.

Consider if your neighbours might benefit from your help.

The Emergency Services will be available on 999 as usual but this should only be used in a genuine emergency as the services are likely to be even busier than usual.

The US government suggests trying to seal windows and doors with tape (Guidelines for staying put[27]) if the air is heavily contaminated. I have not found this suggestion echoed in UK advise.

6.1.3. What can we eat and drink when in shelter?

The people on the nuclear site will be being encouraged not to eat or drink during the emergency in order to minimise the accidental ingestion of any radioactivity that has deposited on the food. This is good advice but should not be carried to extremes and cause discomfort.

Water from the taps should be fine if it comes from a mains water supply. Even if the water supply became contaminated, which is unlikely, it would take a while for the contaminated water to reach your house. But listen to advice on the radio and television. Bottled water will be fine if the bottle has been kept clear of contamination (so a bottle that has been in your house or unopened since the start of the release will be okay). People with their own water supplies are advised to seek advice about their continued fitness for drinking.

---

[27] https://www.ready.gov/shelter

Canned food, packaged food and food from your fridge or freezer is also certainly okay. You might need to use your judgement about food that has been out in the open. This includes food bought from an outdoors market or harvested from an exposed position since the release started.

6.1.4. How long would we be expected to shelter?

It is realised that shelter cannot be held indefinitely. Government guidance considers 72 hours to be a practical limit for no-notice shelter. For some nuclear emergencies, particularly one where a release is prevented, the time in shelter may be only a few hours. In more serious or prolonged releases you may be asked to shelter for longer.

The original NRPB advice on sheltering in the event of a radiological release includes the statement that *"short periods out of doors for necessary activities, will not, in many situations, result in a very high exposure"*[28]. This means you can leave your shelter to check that neighbours are okay or to collect essential medicines should the situation arise. It may be best to seek advice before doing so if possible.

6.1.5. What happens to my children at school?

If your children's school is within a planning zone the school will have a plan to provide shelter in place or will evacuate to a "buddy site" and will care for your child during the school day. What happens at the end of school day is less clear and it may be worth you discussing this with the school when you have the opportunity. You can be assured that your children will be cared for until they can be reunited with you.

If your children's schools are outside the affected area but you live inside the area then it is likely that your children will be kept at school, taken to a Rest Centre or buddy school and be looked after by teachers and local authority staff until they can be reunited with you.

It is probably a good idea to ask your children's schools about their plans for coping with a range of emergencies (fire, flood, illness, road closures etc.).

Government advice on emergency planning for schools can be found on the internet[29].

---

[28] Paragraph 8 of NRPB Board Statement on Emergency Reference Levels
[29] https://www.gov.uk/guidance/emergencies-and-severe-weather-schools-and-early-years-settings

6.1.6. What do I do if I think that I, or one of my children, are contaminated?

"Contaminated" means having some radioactive material on your clothes, skin or hair. While this is present it will be giving a radiation dose to the contaminated people and, to a lesser extent, the people around them. More importantly there is the possibility that the activity will get into the body where it can provide a larger dose over a longer period of time. This is why people who may be contaminated are advised not to eat or drink and to keep their hands away from their face as best they can.

First of all, don't panic. The chances of being contaminated are very low so you probably are not. Even if you were it is likely that the levels of contamination will be too low to have any effect.

If you are worried that you may be contaminated there are some very simple but effective steps that you can take. Consider that contamination is just like dirt and imagine that you or your children have arrived at home very dusty and think about how you would manage that situation.

The most effective form of decontamination is to change your clothes especially your outer layers as this generally removes about 80% of any external contamination. This should be done before walking around to avoid the spread of contamination in the house. Try to peel the clothes off in a way that they end up inside out and the outer surfaces do not touch the skin (this keeps any surface contamination within the clothing). Put the discarded clothes in a bag, a bin bag is ideal, and seal it. A good technique is get the person to try to stand in the bin bag while undressing and try to push the clothes into the bag without the outer surfaces of the clothes touching anything. (Think dirty child on expensive light coloured carpet).

It is a good idea to gently blow the nose and to clean around the nose, eyes, ears and mouth with wet wipes or damp tissue working outwards to avoid pushing any contamination into the eyes, ears, nose or mouth.

If you want to wash then do so with nothing more than soapy water for the body and mild shampoo for the hair. Do not use a hair conditioner since these can fix the radioactivity into the hair. Try to avoid getting runoff water into the mouth, eyes, nose or ears. Do not rub the skin hard. If you make the skin red through rubbing you allow a pathway for contamination to get into the body which only makes things worse.

After washing be sure to thoroughly rinse out the bath, shower or basin used and to bag any cloths or towels used.

If you are still worried you could contact the authorities and see if a Radiation Monitoring Unit (RMU) was being set up in the area. Please do not go to an Accident and Emergency Centre unless advised by a medical practitioner.

Please try to remain calm during this process. Children very easily pick up on parents' and carers' concerns and this can be more damaging than any low levels of radiation or contamination.

A very good infographic on personal decontamination can be found on the US Centers for Disease Control and Prevention web-site[30].

6.2. Taking stable iodine tablets

Some sites in the UK have distributed stable iodine tablets (PITs) to all of the households and businesses in the area surrounding the site out to a distance determined by the regulators. Other sites have plans to distribute the tablets on the day. The prior information provided by the site operator will tell you which plan, if any, applies to you.

If you have been given stable iodine tablets you should only take them when advised by the operator or the police. Tablets should be taken with a glass of water following the instructions given with the tablets.

The instructions to take them will be something like:

- For babies, crush the quarter tablet and then dissolve it in a small quantity of milk or juice – shake well to make sure the power dissolves.
- For young children, crush the half tablet and mix with a teaspoon of jam, honey or yoghurt.
- For older children, crush one tablet and mix with a teaspoon of jam, honey or yoghurt.
- For adults, swallow the two tablets with water; if that is difficult, crush as for children.

But note that the number of tablets might be different if you have been given tablets with a different amount of stable iodine in them.

You only need to take one dose to be protected for a day.

---

[30] https://emergency.cdc.gov/radiation/pdf/Infographic_Decontamination.pdf

When these tablets were given to very large numbers of people in Russia and Poland following the accident at the Chernobyl nuclear power station in 1986, very few people suffered any side effects. The most common one was feeling sick for a while. So, if you feel a bit sick after taking them do not worry, this is not radiation sickness.

If you don't have enough tablets for everyone in your house it is better to give them to any children and pregnant women[31]. This is not based on ancient ideas of chivalry but on the finding that babies and children are more susceptible to harm from radioiodine than are older people. In fact the World Health Organisation (WHO) questions the use of stable iodine on those over 40 as, for them, the risks of thyroid cancer is low[32]. A UK Department of Health working group[33] considered this position and concluded that *It is not necessary to move to a system of age-related ERLs in the UK*".
The tablets protect your thyroid gland from radioactive iodine. They do not protect you from other forms of radiation and contamination so it is important that you remain in shelter as much as possible or evacuate despite having taken the tablets.

More details about the use of stable iodine can be found on the WHO website[34] but note that this describes potassium iodide (KI) whereas the tablets distributed in the UK are potassium iodate (KIO_3). This does not affect how they work but does change the dosages required for protection as less of the UK active ingredient is iodine. Details of UK recommended dosing levels can be found on the internet[35].

If the exposure to inhalation dose continues for a long period then WHO recommends repeat doses of stable iodine daily and claims that there is no evidence that this is any more harmful than one dose. Repeated doses are not the preferred countermeasure for radio-iodine in food as it is better to prevent such contaminated food being eaten.

Potassium Perchlorate is an alternative for people with iodine sensitivity.

---

[31] Importantly stable iodine taken by a pregnant women also protects her unborn child.
[32] http://www.who.int/ionizing_radiation/pub_meet/tech_briefings/potassium_iodide/en/
[33] https://www.gov.uk/government/uploads/system/uploads/attachment_data/file/425072/Documents_of_the_NRPB_Volume_12_Number_3.pdf
[34] http://www.who.int/ionizing_radiation/pub_meet/tech_briefings/potassium_iodide/en/
[35] http://www.england.nhs.uk/wp-content/uploads/2015/04/potassium-iodate-85-adlts-child1.pdf

6.3. Evacuation

If the release of radioactive material is expected to last for a long time and at a level that is dangerous to future health then you might be advised to evacuate. Unless emergency powers have been obtained then the authorities cannot force you to leave your home but only advise.[36]

According to **national guidance**[37] *"The purpose of evacuation is to move people and (where appropriate) other living creatures away from an actual or potential danger to a place that is safer for them"*. The guidance warns that *"Evacuation should not be assumed to be the best option for all risks, and it may not be the safest. Buildings offer significant protection against most risks, and immediate shelter may be safer for the public, at least initially. Evacuation can be traumatic, especially for vulnerable people, and it can be disruptive to business and the local economy"* so evacuation would only be recommended if the release of radioactivity was a serious one that was likely to continue for some time.

Depending on the circumstances different types of evacuation might be ordered.

Dispersal is the kind of evacuation that can occur in public places such as shopping centres, sports venues, theatres or cinemas where the venue is simply closed and the customers requested to leave. This could be used where there is sufficient notice that a release might occur to allow people to manage their own exit. It is likely that stress will be put on local roads and transport systems that planners and responders may be able to alleviate to some extent. In this case you should follow instructions and try to remain patient.

A **Phased Evacuation** is one where the more vulnerable, that is those most at risk due to their location or those whom it will take longer to evacuate, are assisted to evacuate first. It is realised that this may be difficult if a significant number of people decide to self-evacuate prior to being requested to evacuate.

If you are advised to evacuate from your home you should:
- Lock all doors and windows;
- Take any medicines you may need, sufficient for at least three days;
- Take some changes of clothes, being prepared for different temperatures and some overnight stays. Take sleeping bags if you have them;
- Take basic toiletries;
- If you have children then take some things to keep them occupied;

---

[36] Part 2, paragraph 22 of the Civil Contingencies Act states that Emergency made by Act of Parliament or by the exercise of the Royal Prerogative may *require, or enable the requirement of, movement to or from a specified place*; http://www.legislation.gov.uk/ukpga/2004/36/contents

[37] https://www.gov.uk/government/uploads/system/uploads/attachment_data/file/274615/Evacuation_and_Shelter_Guidance_2014.pdf

- Take chargers for your phones and tablets;
- Turn off all electrical and gas appliances and check that no taps are running;
- If you know someone who is elderly, infirm, vulnerable or housebound, tell a member of the Emergency or Social Services.

See, for example, Gosport Borough Council advice[38].

If you are evacuated you have the choice of going to a council Rest Centre where you should find basic support of food and shelter or you could make your own arrangements such as going to friends or family outside the affected area or other arrangements.

6.3.1. What happens at the Rest Centres?

Voluntary organisations and local authority organisations work together to provide space and facilities for displaced people. Rest Centres are selected for their ability to provide suitable spaces to support displaced people with hygiene and feeding being of the highest priority.

On arrival at a centre you will be registered which is a process for determining and recording who has arrived and assessing their needs. In most cases you will be handed a simple form to fill in. Staff will be available to advise if necessary. The completed form is exchanged for an identity card which allows you to use the facilities. You will probably be given a leaflet explaining the facility. Examples of these are available on the internet (Dudley[39], Sussex[40], Gloucester[41]).

It is a good idea to contact friends and relatives to reassure them that you are safe.

At the centre you will be able to get refreshments and, if needed, meals. Areas may be designated for children, for elderly people, for prayer and for pets.

Overnight stays may be required and attempts will be made to provide bedding and privacy but unfortunately it may not be a particularly comfortable experience.

---

[38] https://www.gosport.gov.uk/sections/your-council/emergency-planning/evacuation-advice/
[39] http://www.dudley.gov.uk/resident/your-council/emergencies/emergency-plans/rest-centres/
[40] http://www.sussexemergency.info/glossary-and-faqs/what-can-i-expect-at-a-rest-centre-(rc)
[41] http://www.gloucestershire.gov.uk/media/adobe_acrobat/0/6/Emergency%20Rest%20centre%20Oleaflet.pdf

A Japanese guide to emergency shelters[42] advises that it is important to respect the needs of other people in the shelter. You can help by being patient and calm and helping where possible. You should try to understand the different needs of, for example, mothers with babies and young children, the young children themselves and elderly people. It also recommends particular care of your own health and cleanliness to prevent the spread of infectious disease, food poisoning and vermin - none of which should occur in a well-run facility but all of which should be guarded against.

There are a variety of other centres that might get set up in response to an emergency if they are required (see table).

---

[42]
http://www.metro.tokyo.jp/ENGLISH/GUIDE/BOSAI/FILES/01_Simulation_of_a_Major_Earthqu
ake.pdf

Title	Purpose	When	Lead
Casualty Bureau	Initial point of contact for receiving/assessing information about victims, to: – inform the investigation – trace and identify people – reconcile missing persons – collate accurate information for dissemination to appropriate parties	Immediate If significant number of people killed, hurt or missing	Police
Survivor Reception Centre (SRC)	A secure area in which survivors not requiring acute hospital treatment can be taken for short-term shelter and first aid. Evidence might also be gathered here.	Immediate If significant number of survivors/ walking wounded	A survivor reception centre might be established and run initially by the emergency services – those first on the scene – until the Local Authority becomes engaged in the response, and assumes lead role.
Family and Friends Reception Centre (FFRC)	To help reunite family and friends with survivors – it will provide the capacity to register, interview and provide shelter for family and friends.	12 hours If large numbers of calls to casualty bureau or 'Searching behaviour is noted.	A family and friends reception centre would be established by the police in consultation with the Local Authority, and staffed by these organisations and suitably trained voluntary organisations. Representatives of faith communities might be consulted and interpreters may be required.
Rest Centre	A building designated or taken over by the local authority for temporary accommodation of evacuees/homeless survivors, with overnight facilities.	Overnight If Significant number of displaced people	Lead responsibility sits with the Local Authority, with contributions from police, primary care trusts and the voluntary sector.

Table 2 Immediate sources of help and information[43]

---

[43]

https://www.gov.uk/government/uploads/system/uploads/attachment_data/file/61221/hac_guidance.pdf

6.3.2. What are Humanitarian Assistance Centres?

In the first 24 hours or so the priority will be to provide basic shelter to those displaced from their homes; to ensure that the names and contact details of those displaced are recorded; and to offer a single point of information about what is happening for the local community.

Over a slightly longer period the support offered should become more tailored to the specific needs of the event and the affected community in terms of practical assistance with getting on with their lives and emotional support if it is needed. This may lead to the opening of a Humanitarian Assistance Centre (HAC) or a Community Assistance Centre. This may evolve around a Rest Centre or may be set up in a different facility if that is more suitable. Note that the term "HAC" is a planning term and the facility set up may have a different name appropriate for the situation. Humanitarian assistance can be defined as *"Those activities aimed at addressing the needs of people affected by emergencies; the provision of psychological and social aftercare and support in the short, medium and long term."*

The stated purpose of a HAC is to[44]:

- Act as a focal point for information and assistance to bereaved families and friends of those missing, injured or killed, survivors, and to all those directly affected by, and involved in, the emergency. This group is likely to include the friends and families of those missing and killed, survivors, and the wider community.
- Enable those affected to benefit from appropriate information and assistance in a timely, co-ordinated manner.
- Where necessary, facilitate the gathering of forensic samples in a timely manner, in order to assist the identification process.
- Offer access to – and guidance on – a range of agencies and services – allowing people to make informed choices according to their needs.
- Ensure a seamless multi-agency approach to humanitarian assistance in emergencies that should minimise duplication and avoid gaps.

In the case of a reasonably foreseeable nuclear accident in the UK the first and third bullet points above should not be relevant. In this case the facility might be called a Community Assistance Centre concentrating on practical and emotional support for those whose life was affected by the event.

---

[44]

https://www.london.gov.uk/sites/default/files/gla_migrate_files_destination/London%20HAPlan%20v4.pdf

From the user point of view a HAC is a facility where they can access advice from a range of organisations such as the local authority, benefits, Family Liaison Officers, accommodation advice, transport advice, Citizens' advice, banks, insurance advisors, charities etc.

6.3.3. What are Radiation Monitoring Units (RMUs)?

Figure 3 An RMU set up for an exercise (from guidance document)

According to the government RMU Planning and Operational Guidance[45] an RMU is *"used to determine levels of radioactive contamination in or on people and any subsequent requirement for decontamination"*. It is a facility where you will be carefully monitored for radioactive contamination and given advice based on the findings.

The intent of an RMU is to provide reassurance monitoring to people who fear that they may have been contaminated by a release as well as provide measurements and records of any actual contamination that may have happened. They will be managed by people with suitable medical physics backgrounds and staffed by Public Health England, hospital and operator personnel.

---

[45] https://www.gov.uk/government/uploads/system/uploads/attachment_data/file/340153/HPA-CRCE-017_for_website.pdf

They will generally be set up in sports halls as these provide the space required as well as showers for decontamination. Polythene tracks are put on the floor to guide the users and also to make it easier to remove contamination, should there be any. The "Portals" that you see people walking through contain radiation detectors above, below and either side of the person being monitored. When it is your turn you will be asked to stand in the portal for a few moments and then move on.

As you can see from the picture the Unit's staff are likely to wearing overalls and overshoes. This should not be of concern. They are a precaution against the spread of contamination and allow a quick change of over-clothes if they come in contact with a contaminated person[46].

When you arrive at the RMU you will be registered and asked about where you have been during the event. You will be given a leaflet to read that explains a bit about what will happen to you in the Unit and a little about radioactive contamination and the potential effects it might have (see page 55 of ref[47]).

The first stage of monitoring is a relatively quick process designed to identify people with significant contamination. These will be taken aside and monitored more thoroughly and, if required, attempts will be made to decontaminate them.

During the second stage the monitoring will be slower and more careful, designed to detect lower levels of contamination. If you are shown to have levels of radiation contamination above the "Action Levels" agreed for the event then you will be offered decontamination and re-monitored. Further advice would be available if this happened to you or a member of your family.

If the event involved an operating nuclear reactor then your thyroid gland may be monitored which involves a probe being held to your throat for a while and then to another part of your body. By comparing the readings it is possible to determine if radioactive iodine has concentrated in the thyroid gland.

---

[46] There appears to be a misconception that the coveralls protect the wearer from radiation (quote from 2015 web report ref below reporting the doses to responders *"The new survey comes as a surprise to the public as they were made to believe that the workers assisting in evacuation efforts suffered zero exposure as they were wearing full-body radiation suits and masks"*). The coveralls protect the wearer from becoming contaminated and help prevent the spread of contamination. The face masks reduce the inhalation of contamination. They provide no protection from penetrating radiation such as gamma radiation. https://www.rt.com/news/319902-fukushima-evacuation-radiation-dose/

[47] https://www.gov.uk/government/uploads/system/uploads/attachment_data/file/340153/HPA-CRCE-017_for_website.pdf

It is important to try to stop contamination entering the body so you need to be careful about what you eat and drink and about putting potentially contaminated hands in or near your mouth before you have been monitored and confirmed clean. If you have a wound that is, or has been bleeding, it should be washed carefully ideally using wet wipes or similar brushed away from the wound or by flushing with clean water or saline solution, directing the runoff into a container or drain.

The UK Guidance for RMUs can be found on the internet[48]. An American version can be seen for comparison[49]. A comprehensive description of decontamination techniques from America is also available[50].

6.4. Mass decontamination

In the very improbable event that decontamination of members of the public is required then it is likely to be undertaken in Radiation Monitoring Units (RMUs) or in the people's homes based on advice given.

The UK has further capability to undertake mass decontamination from the CBRNI programme[51]. This provides for *"the continued provision of a system for the decontamination of large numbers of persons who have become contaminated with chemical, biological or radiological substances."* It is unlikely to be needed in a nuclear emergency but worth mentioning.

The service consists of a number of Mobile Response Units operated by the Fire Services which can deploy decontamination facilities that are usually of the "bouncy castle" variety. These are inflatable units that provide undressing areas, showers and dressing areas that large numbers of people can pass through quickly to be decontaminated.

The guidance notes[52] that *"The issues of public disrobing will be difficult for the majority of people and may be traumatic for some. Ensuring high levels of decency is vital"* and hopes that responders will remain *"sensitive to the dignity, cultural and religious concerns and requirements of different communities and social groups and of the special needs of individuals"*. It is however, a situation where dignity might come second to pragmatic medical needs.

---

[48] https://www.gov.uk/government/uploads/system/uploads/attachment_data/file/340153/HPA-CRCE-017_for_website.pdf

[49] https://emergency.cdc.gov/radiation/pdf/operating-public-shelters.pdf

[50] http://www.remm.nlm.gov/ext_contamination.htm

[51] http://www.fireresilience.org.uk/workstreams/national-resilience-capabilites/cbrne-mass-decontamination/

[52] https://www.gov.uk/government/uploads/system/uploads/attachment_data/file/62507/sng-decontamination-people-cbrn.pdf

6.4.1. How might the evacuation be managed?

A major evacuation would probably be managed by the Police Silver (tactical)
commander supported by a multi-agency Evacuation Advisor group (EAG).

You may be told about an **Evacuee Assembly Point.** This is the name given to one
or a number of locations for evacuees to meet and be signposted to safety. These
are usually identified on the day. The Police will make the decision based on the
likelihood of an evacuation.   They will consider the location of the incident and
the nature of the Incident Cordons.

Evacuees who are unable to 'self-evacuate' may be directed to **Evacuee Transit
Points** where they will receive advice and services to enable them to move on to
family and friends or to alternative accommodation.

It is important that you stay in shelter while the evacuation is being organised. The
radiation dose you'll receive waiting in your house is likely to be significantly less
than if you leave your house and then have to wait out in the open for evacuation
instructions and transport or if you get stuck in a traffic jam.

6.4.2. What happens to my pets?

If you have pets you need to consider how best to care for them given that you
don't really know how long you will be away from your home. This is something
you can think about before the event (see the leaflets by Greater Manchester[53] or by
Somerset CC[54] or South Gloucestershire's website[55] for some ideas). If you had to
leave your home how would you transport your pets? And do you have friends or
family that could look after the pets for you for a few days?

The Borough of Poole council give the following advice on their web-site[56]: *If you
have to leave your pets behind ensure you do the following:*

- Do not tether/tie them;
- Leave them in a room with plenty of free-space, upstairs if a flood risk.
- Never leave them if risk is fire;

53

http://www.gmemergencyplanning.org.uk/gmprepared/info/12/protecting_your_pet_during_an_
emergency

[54] http://www.somerset.gov.uk/EasySiteWeb/GatewayLink.aspx?alId=42846

[55] http://www.southglos.gov.uk/health-and-social-care/staying-healthy/health-
protection/emergency-planning-and-weather-advice/emergencies/preparing-for-emergencies/pets-
in-rest-centres/

[56] http://www.poole.gov.uk/environment/animal-welfare/pets-and-evacuations/#What-if-I-have-
to-leave-my-pet-behind

- Leave enough food and water for 3 days (per animal);
- Ensure your pet is wearing identification (where appropriate). Having the pet microchipped before the event and having a note of the chip number in your emergency pack is a good idea;
- Leave a note on your front door, or somewhere immediately visible, for the emergency services advising them there are animals in the house, what those animals are and where they are.

If you cannot find your pet before evacuating you should try to find and complete a *missing pets form*, these may be available at Rest Centres. These are passed to the police who may be able to reunite pets and owners.

If you take a pet to the Rest Centre they will try to accommodate it, either with you or in separate facilities. But you must remember that there may be many other people displaced and their needs must be considered. If you arrive by car then one option may be to leave your pets in the car taking care to ensure their welfare with respect to temperature, food and water.

6.4.3. What happens to farm animals?

Ideally farm animals would be removed from the fields affected by the radioactive cloud. Farmers are often advised to get the animals inside if possible and to feed them with foods stored before the release started. The animals being indoors and not eating contaminated foods is the best way to avoid uptake of contamination and to avoid passing radioactivity along the food chain.

After the event the animals would be monitored and a decision made about their future. It is quite likely that any milk produced in affected areas will be unsuitable for human consumption due to its contamination with iodine and caesium isotopes. The sale of milk was banned around Sellafield following the Windscale Fire in 1957 and the sale of sheep was controlled in Wales and Cumbria for some time following the Chernobyl accident.

Advice from Somerset County Council on the protection of animals can be found on the internet for livestock[57] and for horses and ponies[58].

---

[57] http://www.somerset.gov.uk/EasySiteWeb/GatewayLink.aspx?alId=42850
[58] http://www.somerset.gov.uk/EasySiteWeb/GatewayLink.aspx?alId=42845

6.4.4. Return from Evacuation

A nuclear emergency is likely to be different to anything experienced in the UK. After an evacuation initiated by a flood or a storm there is often a great deal of debris to clear from roads and obvious damage to homes and other buildings that needs to be cleaned up. Making the area safe involves draining any water, clearing debris, re-establishing the essential services, drying, cleaning and repairing buildings. See an American web-site on the subject[59]. This can be a heart breaking, time consuming and expensive process.

For a nuclear emergency the streets, gardens and houses are likely to look undamaged but those downwind of the accident could be contaminated with radioactive dust. This contamination will be invisible but can be detected at very low levels using a range of instruments. If left in the environment it will produce an additional radiation dose to anyone in the area and contaminate food grown in the area.

During and shortly after the accident the dispersion of activity from the site will be modelled to suggest where radioactivity might be deposited and teams will be sent out to measure radiation levels and contamination levels. These data will be used to determine those areas that are unaffected and to which the occupants can return; those areas which are contaminated but at a low level so occupants can return after being informed of the situation and those areas to which it would be unwise to return other than brief visits to ensure the safety of the property and collect essentials. Unless it is a severe accident the number of people affected by significant levels of contamination is likely to be small.

Following a release of radioactive material, the site levels of deposition can vary considerably resulting in areas with higher concentrations of radioactive material or dose rates than those nearby. These are sometimes called "hot-spots". These are expected to occur where the wind is interrupted by, for example, trees, hedges or hills and where drainage systems deposit radioactive dusts from roads, roofs and other hard surfaces. See Figure 4.

---

[59] http://www.floodsafety.com/national/property/cleanup/

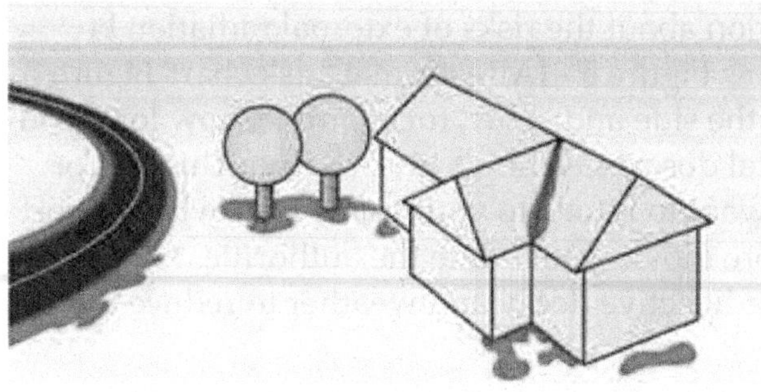

Figure 4 Showing where "hot-spots" might be expected

From IAEA EPR-NPP Public Protective Actions 2013[60]

Over the next days, months and years further measurements will be made of the levels of contamination in the environment and consideration will be given to reducing it where it is practical and cost-effective. A number of things can be done such as spray washing roads and walls, cutting grass and foliage and disposing of it and so on as detailed in a series of PHE documents[61]. This is an expensive and time consuming process that can cause a lot of environmental damage[62]. There will be a debate about the targets for how far to reduce dose. One of these documents[63] makes the point that *"Unlike emergency situations where prompt response toward preserving life and critical infrastructures is the overriding consideration, more time is available in the recovery phase to develop effective schemes for involving stakeholders. Recovery is necessarily community focused and community-based and stakeholders representing local needs can provide essential input on the complex and multi-faceted issues facing the recovery programme"* suggesting a continuing discussion between those managing the clean-up and those living, working or otherwise using the affected areas.

If you return to an area that is contaminated you will be given advice about behaviours that might affect your radiation dose. You may be advised to avoid eating locally produced foods, particularly wild foods such as mushrooms.

---

[60] http://www-pub.iaea.org/MTCD/Publications/PDF/EPR-NPP_PPA_web.pdf

[61] https://www.gov.uk/government/publications/uk-recovery-handbooks-for-radiation-incidents-2015

[62] http://e360.yale.edu/feature/as_fukushima_cleanup_begins_long-term_impacts_are_weighed/2482/

[63]

https://www.gov.uk/government/uploads/system/uploads/attachment_data/file/432742/PHE-CRCE-018_Inhabited_Areas_Handbook_2015.pdf

The IAEA recommend that information about the risks of external radiation be given in the form of risk charts such as Figure 5 - IAEA format "risk chart"Figure 5. These show increasing dose-rate up the side and show, for example, how long you can live in any area without your total dose exceeding a given value. This sort of presentation might be useful if you want to return to your home for a while to sort things out and collect valuables before moving out while the authorities decontaminate the area or wait for radioactive decay and weather to reduce the dose-rates to more acceptable levels.

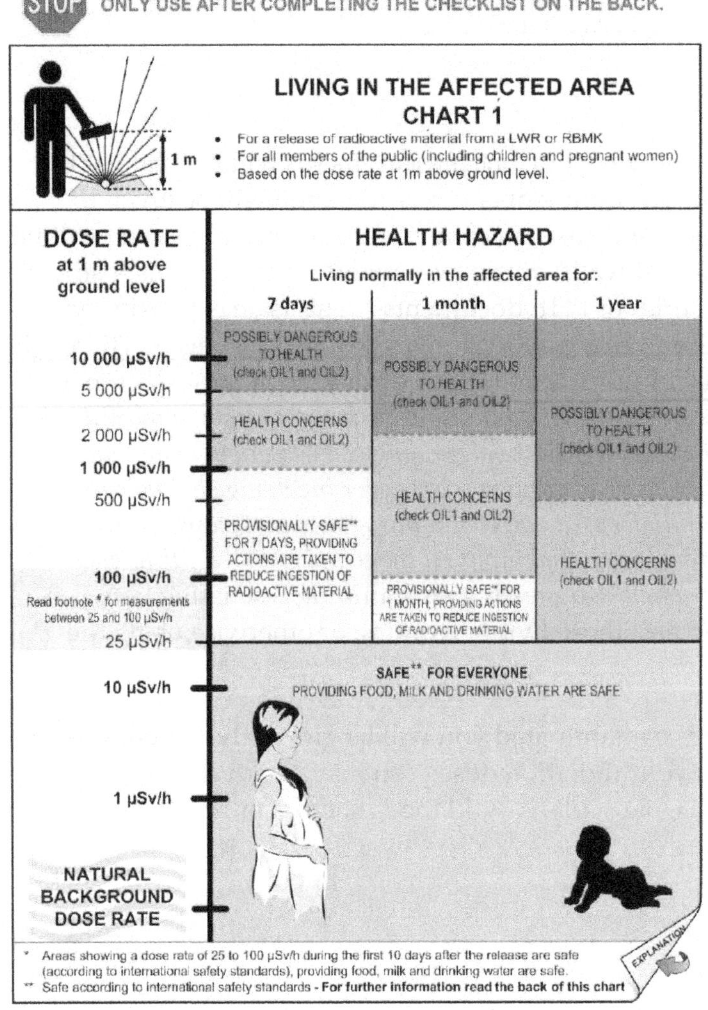

Figure 5 - IAEA format "risk chart"
From EPR NNP Public Protective Actions[64]

---

[64] http://www-pub.iaea.org/MTCD/Publications/PDF/EPR-NPP_PPA_web.pdf

At this stage the physical and mental health of those affected becomes of paramount concern and it would be expected that people would be offered radiation monitoring, health screening and counselling should they wish it.

Faced with these decisions larges areas around Chernobyl were permanently evacuated and the populations relocated. By contrast around Fukushima they tried to minimise depopulated areas and undertook, and are still pursuing, a programme of aggressive decontamination.

7. Caring for yourselves and your family after the event

In summary, if there is a nuclear accident near you your priorities are to stay safe and protect your family, friends and neighbours. You do not want to have an accident of your own so try to stay calm and behave sensibly. Listen to the advice from the police and local authority which will be on the local radio channel and on TV.

In terms of reducing any potential radiation dose you want to avoid breathing in any of the radioactive material which would drift downwind of the site. This means get away from the plume if possible (evacuation) or stay inside a building with doors and windows closed and no heating or air conditioning running (shelter). If it is a reactor accident you might be advised to take stable iodine tablets (PITs). Follow the instructions given.

After the initial phase, when the leak has stopped and the plume has gone by, the aim is to avoid eating any of the radioactive material. This means being careful about the food and drink you consume. Could it have got contaminated?

Only in a very big accident would contamination of the environment cause a major problem after the accident is over. Post-accident there will be a lot of monitoring taking place and a programme to decontaminate heavily affected areas would be discussed. There may be advice given about changes in behaviour that would help reduce your dose. This might include advice not to eat local wild foods such as berries and mushrooms, to avoid spending too long in some areas and to avoid some areas entirely. After severe accidents these restrictions could last for many years.

It is expected that this stage will be highly interactive with a series of meetings for the local communities to review any residual radioactivity in the environment and to discuss the needs for further monitoring, decontamination or advice.

Involvement in a nuclear emergency would be a traumatic experience for many and continuing to live or work in a mildly contaminated area or undertaking a hasty move to another area can add significantly to the trauma.

There are many reports of people suffering symptoms such as lack of sleep, loss of appetite, panic attacks and poor concentration for a period of time after such an event. For many these symptoms fade with time but some develop post-traumatic stress disorder[65] which can affect their quality of life. If you feel that you, a member of your family or a friend are failing to cope well then you should seek advice as soon as possible. Your Doctor is a good place to start as they can help and they can also refer people to specialist care. Advice is also available from the British Red Cross[66].

Your local authority may have set up a Humanitarian Assistance Centre as a single point of contact to access a wide range of advice and support services. There may be police Family Liaison Officers offering support and there may be self-help communities set up to keep you in touch with people who have had similar experiences as this has be shown to help some people.

For Chernobyl and Fukushima it is not clear if the stress symptoms reported resulted from memories of the day itself, fear about the impact of any addition radiation dose they may be subject to or are a result of the changes resulting from the accident. It is likely to be a combination of these with different people reacting differently.

Following Fukushima and Chernobyl many people were unable to continue living in their old homes and had to move to new areas in different communities where many struggled to settle in. Others were able to return to live in their homes but were aware that the area was contaminated with radioactive material so that they, and their family, were exposed to higher levels of radiation than before.

For an accident in the UK it is unlikely that a wide area of land will be significantly contaminated. In that event Public Health England, the environment agencies and the local authorities will work together in the Recovery Working Group. National Guidance for this group can be found on the internet[67].

---

[65] https://www.nhs.uk/conditions/post-traumatic-stress-disorder-ptsd/
[66]
http://www.redcross.org.uk/~/media/BritishRedCross/Documents/Archive/GeneralContent/C/
Coping%20in%20a%20crisis.pdf
[67] https://www.gov.uk/guidance/national-recovery-guidance

8. Summary

This book has tried to explain the countermeasures that might be advised if there were to be a nuclear accident in the UK. These countermeasures of shelter, evacuation and the taking of stable iodine are all designed to reduce the dose to affected members of the public in a way that balances the benefits of the dose reduction with the costs and inconveniences of following the advice.

Emergency plans exist at all nuclear sites within the UK and for those that merit it these include a local authority off-site plan established to ensure the timely imposition of these countermeasures.

By following the advice from the authorities and by staying calm you can reduce the radiation dose received by yourself, your family and your friends. This can be further improved by having an agreed family emergency plan that could cover a wider range of events.

After the event you may be concerned about the effects of any additional radiation dose you, your family or friends might have received during or after the event and your life may be disrupted by continuing controls. This can be very stressful and it is important to be aware of the symptoms of stress and to seek help sooner rather than later should they appear.

Finally, it is considered very unlikely that there will ever be a serious nuclear accident in the UK but there is no room for complacency we need the operators to be systematic and careful in the way they work and the regulators to be vigilant in their oversight of the industry. Where appropriate emergency plans should be produced, tested and revised and the public who might be affected informed, prepared and ready to trust the operators and authorities when they give advice.

Appendix A: Nuclear Reactors, Accidents and Emergency Plans

A.1   How nuclear reactors work

Nuclear reactors work by breaking large *nuclei* (the middle bits of atoms) into smaller ones and in doing so liberating some of the energy bound within the atomic nucleus. At the heart of all current nuclear reactors is the *chain reaction* in which a neutron hits a large nucleus (usually U-235), causing it to split (the technical term is *fission*) into a number of bits including smaller nuclei (*fission products*) and some new neutrons. If some of these neutrons can be made to hit other heavy atoms and start the process again you get a chain reaction which keeps going.

The fission process produces a lot of heat which is used in nuclear power plants to make steam to drive turbines and generate electricity. It also produces *fission products* or fission *nuclei* which are smaller nuclei. These fission products are usually *unstable* and therefore *radioactive*. That means that they will *decay* at some time in the future giving off nuclear radiation and more heat. This causes two problems that have to be managed by the design and operation of the reactor; (1) the radiation is harmful so the fission products must be kept safely behind shields away from people and the environment and (2) the decay heat must be removed from the reactor core even after the reactor has shut down or the system will heat up and could damage itself.

In nuclear reactors the heat produced by fission and decay heat is removed by pumping a *coolant* through the reactor which takes the heat away. In some reactors such as Boiling Water Reactors this coolant is water which boils above the reactor to make steam to drive turbines. In other designs, such as the Pressurised Water Reactor (PWR) there is a secondary circuit which removes the heat from the reactor coolant to make steam to drive the turbines. In the British Advanced Gas Cooled Reactors (AGRs) the coolant is carbon dioxide gas.

When the reactor is shut down the fission products continue to decay and so continue to produce heat. This quickly reduces so a shutdown reactor requires far less cooling than an operating one. However, if there is not enough cooling the system can heat up and can damage itself. That was the situation seen at Fukushima where the tsunami damaged the reactor cooling systems and their backups. This allowed temperatures to rise sufficiently to damage fuel elements and reactor cores and to produce hydrogen from metal/steam reactions. It was hydrogen explosions that provided the more dramatic TV pictures.

A.2   A brief history of nuclear accidents

There have been a number of nuclear accidents over the years and several of these
have been big news items for a long period of time. Most people who are old
enough remember the disruption to lives caused by the Chernobyl (1986) and
Fukushima accidents (2011) and many people worry that it could happen to them.
The headlines make uncomfortable reading, reporting thousands of people fleeing
their homes, some never to return to them. But could it happen in the UK and
what impact would it have on our lives?

The IAEA have analysed the lessons learned from a large number of accidents that
occurred between 1945 and 2010[68] but this section will concentrate on reactor
accidents.

The Windscale Fire

The first nuclear accident that many people in the UK would have been aware of
was the Windscale Fire in October 1957. The two Windscale piles had been built in
a hurry to produce material for the nuclear weapons programme and were a
remarkable achievement given the resources and knowledge then available.
However, they were working at the limits of both technology and knowledge and
on the 8th October a process on Pile No. 1 went badly wrong and the graphite core
caught fire. The fire was to burn for three days.

In those days there was no emergency planning in the sense we understand today
but the enquiry held after the event found that steps taken to deal with the
accident, once discovered, were *'prompt and efficient and displayed considerable
devotion to duty on the part of all concerned'* (Penney report quoted by Wikipedia[69]) It
was realised that radioactive iodine was being released and that this could enter
the food supply through locally produced milk. As a consequence of this milk
produced in the area in the following month was destroyed. A month was
considered to be adequate as the isotope of concern was I-131 with a *half-life* of
about 8 days. After a month this isotope had largely decayed away. After 4 half-
lives there would be 1/16th of the original quantity left.

The transcript of the public enquiry is available on the internet[70].

---

[68]   https://www-pub.iaea.org/books/IAEABooks/8920/Lessons-Learned-from-the-Response-to-
Radiation-Emergencies-1945-2010
[69] https://en.wikipedia.org/wiki/Windscale_fire
[70] http://news.bbc.co.uk/1/shared/bsp/hi/pdfs/05_10_07_ukaea.pdf

---

The accident at Three Mile Island

In the early morning of March 28, 1979 the pumps pushing water into the heat exchangers at Three Mile Island Reactor 2(TMI-2) failed causing the turbine-generator and then the reactor to shut down. When nuclear reactors shutdown the chain reaction, which generates most of the heat, ceased but the decay of fission products continued to produce *decay heat*. With the reactor cooling switched off temperatures and pressures increased rapidly and a pressure relief valve at the top of the pressuriser opened as intended to release the pressure. This valve then failed to reseat (close) once the pressure had reduced as it should have done and a failure in the instrument panel hid this fact from the operators. There followed a sequence of events that led to the reactor core melting but there was little release of radioactive material as the reactor was surrounded by a *containment vessel.*

A general emergency, defined as having the "potential for serious radiological consequences to the general public" was declared at about 06.30. Over the next few days there was confusion about the extent of reactor damage and the possible consequences. Schools were closed, residents were advised to stay indoors and farmers told to keep animals under cover and fed with stored feed. The local Governor advised evacuation of pregnant women and pre-school children within 5 km of the site and later this distance was increased to 20 km. Within days 140,000 people had left the area (voluntary, self-evacuation) almost all of whom returned to their homes within three weeks.

Despite the usual scare stories and non-peer reviewed papers claiming health effects in both people and animals in the area all reputable sources claim that the health effects of the TMI-2 accident were unmeasurably small.

More details about the reactor and the accident can be found on the internet[71].

Chernobyl

The Chernobyl accident happened on 26 April 1986. This was a soviet designed RBMK-1000 which had a number of design features that made it dangerous to operate in certain ways. On the day of the accident the reactor was the subject of an experiment, several safety systems were switched off and the reactor was being operated outside its normal operating regime.

Two large explosions wrecked the reactor building and then the fuel over heated and caught fire in several places. Two workers died in the explosions.

---

[71] See, for example:
http://www.nrc.gov/reading-rm/doc-collections/fact-sheets/3mile-isle.html, **or**
http://www.world-nuclear.org/information-library/safety-and-security/safety-of-plants/three-mile-island-accident.aspx

Water was poured into the core to cool it but this was stopped after a while as there was a risk of flooding the other reactors on the site. From the second to tenth day after the accident, some 5000 tonnes of boron, dolomite, sand, clay and lead were dropped on to the burning core by helicopter in an effort to extinguish the blaze and limit the release of radioactive particles.

The accident deposited radioactive dust and debris over Europe but particularly over Belarus, Russia and Ukraine. Iodine-131 was the isotope of greatest concern in the early days following the accident with Cs-137 taking over when the short lived Iodine had decayed.

The cloud of radioactive material reached as far as the UK and, while it was over Cumbria and North Wales it rained heavily washing radioactive caesium (Cs-137 and Cs-134) out of the plume and onto the ground. On most soils the caesium would have soon become "fixed" into the soil and not available for uptake by crops but not on the acid soils found in the Cumbrian, Scottish and Welsh uplands where the activity continued to be found in grass and the sheep that grazed that grass for many years following the deposition.

The Food Standards Agency[72] monitored the levels of radioactivity in grass and sheep and imposed a limit on the acceptable concentration of radio-caesium in sheep (1000 Bq/kg). Meat above this level could not be sold. Some controls lasted until 2012[73].

According to a WHO report[74] about 530,000 people were used in the clean up the area around Chernobyl. These received an average additional dose of about 120 mSv. Those evacuated from the area (115,000 people) received an average additional dose of 30 mSv. This however, is a measure of whole body dose. Many people in the area received high doses to the thyroid gland due to the release of radio-iodine which was breathed in as the plume passed but, more importantly, entered into the food chain particularly milk over the following weeks. A few people continue to live in areas where the radiation dose is double the background levels experienced before the accident.

28 liquidators died of acute radiation sickness (ARS) within three months of the accident. There was a significant increase in the levels of thyroid cancers, mainly in children, in the areas around Chernobyl. An increase in the incidence of leukaemia in the most highly exposed liquidators has also been seen.

---

[72] http://www.food.gov.uk/the-website-of-the-food-standards-agency
[73] http://www.food.gov.uk/science/research/radiologicalresearch/radiosurv/chernobyl
[74] http://www.who.int/ionizing_radiation/chernobyl/20110423_FAQs_Chernobyl.pdf?ua=1

A large number of people were moved from their homes never to return in the aftermath of the accident (Figure 6). The social and economic disruption caused had a long term impact on mental health not helped by the survivors being shunned by the residents of areas they were moved into. It is widely believed that the psychological harm caused by the response to the accident are far greater than those caused by the radiation. An important lesson for those preparing against any possible future accidents.

Figure 6 Part of the abandoned town of Pripyat

NRPB have estimated that the average total UK dose from Chernobyl fallout was 0.05 mSv, although the dose to people living in areas of high deposition was higher (e.g., the average additional dose per person in Cumbria was 6 times the UK average). A parliamentary briefing on the impact of Chernobyl on the UK can be found on the internet[75].

---

[75]http://researchbriefings.files.parliament.uk/documents/POST-PN-45/POST-PN-45.pdf

Fukushima

On 11th March 2011 a large earthquake off the eastern coast of Japan caused considerable damage in the region. It unleashed two large tsunamis that inundated an area of about 560 sq.km and resulted in a death toll of over 19,000 people. Over a million buildings were damaged.

The most memorable buildings to be damaged were those of the Fukushima Daiichi nuclear power plant. Here the cutting of electrical supplies, inundation of the emergency diesel and other widespread damage was to lead to a loss of cooling accident and the melting of three reactor cores.

Worldwide media interest in the earthquake, tsunami and nuclear accident was intense but soon focused mainly on the on-going nuclear emergency. The Daily Mail in the UK, for example, ran the head line "A nation in the grip of nuclear panic".

The public around the site were advised to evacuate with the evacuation distance increasing with time extending 20 km by the evening of the 12th March. Beyond that those who might have been exposed to doses in excess of 20 mSv were also advised to evacuate. Determining who to evacuate proved difficult for a number of reasons including the damage that had been done to the radiation monitoring equipment by the earthquake and tsunami.

There was significant criticism of the authorities for the way the emergency was handled. This included the way in which countermeasures were applied.

The accident was rated 7 on the INES scale, due to high radioactive releases over days 4 to 6, eventually a total of some 940 PBq (I-131 equivalent) were released. Radiological surveys found that radioactivity was unevenly distributed with the highest levels in a line to the north west of the site. (See the results of aerial surveys in a World Nuclear Association report[76].)

There are a number of reports that claim that there was no obvious radiation harm either to the workers on-site who recovered the situation or to members of the public evacuated from the area but that a lot of people suffered health effects resulting from the evacuation and, in particular, the stress of the following years.

The UK dose resulting from Fukushima was estimated by Public Health England to be about 1/10,000 of the average annual dose (J.Radiol. Prot. 34 (2014) N41).

---

[76] http://www.world-nuclear.org/info/Safety-and-Security/Safety-of-Plants/Fukushima-Accident/

International Nuclear Event Scale (INES)

The International Nuclear and Radiological Event Scale (INES)[77] is a tool for communicating the safety significance of nuclear and radiological events to the public. Member States use INES on a voluntary basis to rate and communicate events that occur within their territory.

Events are rated at seven levels. The scale is logarithmic – that is, the severity of an event is about ten times greater for each increase in level of the scale. The scale is quite complex to use since it considers impact on people and the environment; impact on radiological barriers and control and impact on defence in depth.

In this scale a level 5 event is one "with wider consequences" i.e. off-site consequences that might involve a limited release of radioactivity and the need to advice some countermeasures. This level could include several deaths from radiation. On-site the situation would probably involve core damage (for a reactor facility) and significant failure of one or more radioactivity containment barrier. The Windscale Fire and Three Mile Island both fit into this category.

A level 6 event involves a "significant release of radioactivity" leading to countermeasures. This would have applied to the Kyshtym disaster at Mayak[78] in the USSR in 1957 which lead to the eventual evacuation of several villages.

Level 7 is the highest level on the scale. It involves a major release of radioactive material with widespread health and environmental effects requiring implementation of planned and extended countermeasures. Both Chernobyl and Fukushima are in this category.

---

[77]   https://www.iaea.org/topics/emergency-preparedness-and-response-epr/international-nuclear-radiological-event-scale-ines

[78]

https://ipfs.io/ipfs/QmXoypizjW3WknFiJnKLwHCnL72vedxjQkDDP1mXWo6uco/wiki/Kyshtym_disaster.html

Description and INES Level	People and Environment	*Radiological Barriers and Controls at facilities*	Defence-in-Depth
Major Accident Level 7	Chernobyl, 1986 — Widespread health and environmental effects. External release of a significant fraction of reactor core inventory.		
Serious Accident Level 6	Kyshtym, Russia, 1957 — Significant release of radioactive material to the environment from explosion of a high activity waste tank.		
Accident with Wider Consequences Level 5	Windscale Pile, UK, 1957 — Release of radioactive material to the environment following a fire in a reactor core.	*Three Mile Island, USA, 1979 — Severe damage to the reactor core.*	
Accident with Local Consequences Level 4	Tokaimura, Japan, 1999 — Fatal overexposures of workers following a criticality event at a nuclear facility.	*Saint Laurent des Eaux, France, 1980 — Melting of one channel of fuel in the reactor with no release outside the site.*	
Serious Incident Level 3		*Sellafield, UK, 2005 — Release of large quantity of radioactive material, contained within the installation.*	Vandellos, Spain, 1989 — Near accident caused by fire resulting in loss of safety systems at the nuclear power station.
Incident Level 2	Atucha, Argentina, 2005 — Overexposure of a worker at a power reactor exceeding the annual limit.	*Cadarache, France, 1993 — Spread of contamination to an area not expected by design.*	Forsmark, Sweden, 2006 — Degraded safety functions for common cause failure in the emergency power supply system at nuclear power plant.

A.3   Can it happen here?

The UK nuclear power industry has a proud history of safe operation due to the high standards of design, construction, operation and maintenance and the close supervision of the Office for Nuclear Regulation[79]. The National Risk Register of Civil Emergencies[80] states that *"Nuclear sites are designed, built and operated so that the chance of an accident and any release is very low"* but notes that *"Accidents have occurred, however, notably at Windscale (UK) in 1957, Three Mile Island (US) in 1979, Chernobyl (Ukraine) in 1986 and Fukushima (Japan) in 2011. In the last two cases, radioactivity was released from the site"*.

The ONR state that "the risk of dying as a result of a nuclear accident in the UK is about 1 in 100 million per year" (taken from Reducing Risks, Protecting People)[81] and compare this to all cancers 1 in 387, all types of accidents and other external causes 1 in 4,064, road accidents 1 in 16,800 and lightening 1 in 18,700,000.

The chances of an unplanned release of radioactivity affecting the public are very small but, nonetheless, emergency plans are written, tested and revised for all nuclear licensed sites in the UK in line with the requirements of the nuclear site license and, for higher risk sites, REPPIR. If you live near a nuclear power station you should understand what might happen in an emergency. The UK currently has **36 nuclear licensed sites**[82] although several of them are in the late stages of decommissioning and probably unable to have an off-site nuclear emergency.

A.4   Reference Accident sequences in the UK

Careful design and operation of nuclear facilities ensured by close regulatory scrutiny is intended to reduce the probability of nuclear accidents to a very low level and reduce the consequences of any such accident should it occur. However, it is realized that rare combinations of events could lead to a release of radioactivity sufficient to affect the public from some of the nuclear sites in the UK.

---

[79] http://www.onr.org.uk/

[80]

https://www.gov.uk/government/uploads/system/uploads/attachment_data/file/419549/201503 31_2015-NRR-WA_Final.pdf

[81] http://www.hse.gov.uk/risk/theory/r2p2.pdf

[82]http://www.onr.org.uk/licensees/pubregister.pdf

All nuclear sites produce a Hazard Identification and Risk Evaluation (HIRE) document every three years of operation. This reviews the site's safety case and determines if a REPPIR Emergency (an event that could give a dose of 5 mSv to a member of the public) is "reasonably foreseeable" which is a legal term that is hard to define precisely. **Guidance** states that *"in the context of a radiation emergency, a reasonably foreseeable event would be one which was less than likely but realistically possible"*[83].

For the gas cooled AGR reactors a significant atmospheric release would require coincident fuel and pressure vessel damage. In this combination of circumstances radioactivity could be released into the coolant from the fuel and hence into the atmosphere as the reactor depressurised. The operators would be taking steps to control the depressurisation and reduce the discharge. The release would largely stop when the reactor is depressurised, at which point attempts would be made to seal the leak. How long the depressurisation takes depends on the leak size and operator actions but is likely to be several hours. Thus the reference accident gives little or no notice and lasts a few hours and, estimates suggest, releases sufficient radioactive material to make countermeasures worthwhile out to no more than about 1 km downwind. It is expected that the release would terminate at this stage but if reactor cooling fails and cannot be resumed then a further release could occur after about a day and require countermeasures to be applied further out.

For nuclear submarines the reference accident relates to a badly damaged reactor and a coincident failure to maintain containment in the reactor compartment and submarine hull. This unlikely combination (probability about 1 in a million reactor years) would give a warning time of the few hours and then last for a few hours until the submarine was depressurised. More severe accidents would entail a greater release rate. They would give the same warning period but would terminate more quickly. For the reasonable foreseeable faults the countermeasures may apply for a kilometre downwind.

For Sellafield the scale of the emergency scheme is based on a reference accident in which an earthquake damages a number of buildings leading to a release of radioactivity.

---

[83] http://books.hse.gov.uk/hse/public/saleproduct.jsf?catalogueCode=9780717622405

A.5 Emergency plans

For sites where an off-site nuclear emergency is *reasonably foreseeable* there is a legal requirement under the Radiation (Emergency Preparedness and Public Information) Regulations (REPPIR 2001)[84] for both an on-site (Operator's) plan and an off-site (Local Authority) plan which should explain how the site will be made safe and how the public will be protected if something goes badly wrong. National Guidance on nuclear emergency planning[85] is published by BEIS.

The Operator's on-site emergency plans are designed to "secure, so far as is reasonably practicable, the restriction of exposure to ionizing radiation and the health and safety of persons who may be affected by such reasonably foreseeable emergencies as are identified by the said assessment" (REPPIR Regulation 7), that is, stop the release and protect those that might be affected. The local authority's off-site plan is similarly designed to protect people within the Detailed Emergency Planning Zone as defined by the ONR. (REPPIR Regulation 9).

The Operator's on-site plan will define a number of sets of circumstances under which it is necessary to declare an off-site nuclear emergency and explain how they will raise the alarm. This generally involves urgently contacting a number of organisations including the emergency services, the local authority, the regulators and central government and informing them of the situation. Each organization will then set up their emergency response organization and will send representatives to local and/or national response facilities as required.

The key response facilities and their roles are summarized in the figure below. This is based on the UK generating companies' emergency arrangements for sites in England and Wales. Different but similar on-site arrangements apply for other organisations and, for sites in Scotland the Scottish Government is involved.

There are a number of facilities on-site to enable the management of the situation and the protection of those responding to the incident and those on-site but without a role in the response. These include the Emergency Control Centre (ECC) from which the site's response is managed and co-ordinated, the Access Control Point (ACP) which manages access to the affected areas, and muster stations, which are safe havens for the personnel not directly involved in the response.

---

[84] http://www.hse.gov.uk/radiation/ionising/reppir.htm
[85] https://www.gov.uk/government/publications/national-nuclear-emergency-planning-and-response-guidance

Off-site the local authority manage the Strategic Coordination Centre (SCC) which is a facility used to manage the local response to any type of incident serious enough to require it. It is at the SCC that the major decisions about what countermeasures to advice to the public are made by a grouping of local leaders and technical support.

Central government would play a key role with the Department for Business, Energy and Industrial Strategy (BEIS), playing an information and coordination role as Lead Government Department and the Cabinet Office Briefing Rooms (COBR)[86] being used to inform government and manage their input to the response. In Scotland, the Scottish Government would be involved based at the Scottish Government Resilience Room (SgoRR)[87].

The Government's default objectives for any response strategy for a radiation emergency are to[88]:

- Protect human life and, as far as possible, property and the environment;
- Alleviate suffering;
- Support the continuity of everyday activity and the restoration of disrupted services at the earliest opportunity; and
- Uphold the rule of law and the democratic process.

---

[86]
https://www.gov.uk/government/uploads/system/uploads/attachment_data/file/192425/CONOPs_incl_revised_chapter_24_Apr-13.pdf
[87] http://www.gov.scot/Resource/0038/00389881.pdf
[88]
https://www.gov.uk/government/uploads/system/uploads/attachment_data/file/472419/NEPRG00_-_Concept_of_Operations.pdf

ACCIDENT SCENE
Boundary of inner area. Coordination of forward responders, contamination control, BA control
Forward Control Point
Facility which controls the personnel entering and leaving the affected area
Access Control Point
Facility from which the Reactors are controlled.
Central (Reactor) Control Room
Facilities and processes ensuring that the site and the material contained there-in remain safe.
Safe haven for people on-site not involved in the response
Site Muster Point(s)
Emergency Control Centre
Site Security Control
Mobile radiological monitoring capability
Location from which strategic control of the affected site is effected
Company Technical Support and management of off-site situation
Company off-site Centre
Off-site Survey
Key local decision making centre (countermeasures to protect the public)
Lead Government Department's coordination of response
Redgrave Court Incident Suite
Strategic Coordination Centre
Nuclear Emergency Briefing Room
ONR coordination
On-site facility
Off-site facility
External facility
Coordination of media briefing
Media Briefing Centre
Cabinet Office Briefing Rooms
Central Government (including PM) involvement coordinated from here
© Katmal Limited 2015

Appendix B: Radiation and Radiation Protection

B.1.   What is radiation and radiation dose?

Public Health England material on this subject can be found on the internet[89].

Some materials contain *unstable atoms* that can undergo spontaneous changes that result in them changing their chemical form and giving out either a particle or wave. These materials are called **radioactive materials** or **radioactive sources** and the particles or waves are called **radiation**. Nuclear reactors produce a lot of radioactive material (fission products and activation products).

The amount of radioactivity is measured as the *Activity* and is expressed in units of becquerel (Bq). A one becquerel source is one undergoing one decay per second. This is a very weak source. More often sources are given with SI prefixes[90] (1 kBq = 1,000Bq, 1 MBq = 1,000,000 Bq; 1 GBq = 1,000,000,000 Bq; 1 TBq = 1,000,000,000,000 Bq).

There are four main types of radiation that we are concerned with when discussing the nuclear industry. These are called alpha (α), beta (β), gamma (γ) and neutron. They are all *Ionising Radiations* (ones that can ionise material they pass through by knocking electrons out of the atoms) but they are very different to each other. See Health Physics Society web-page[91] for more detail.

These radiations interact with the material they are passing through in different ways and the effect can depend on the energy of the radiation, the amount of radiation, what it interacts with, and, for living tissue, how it is changed and the impact that has on the health of the person affected. It quickly becomes rather complicated to explain.

In brief, these radiations interact with material and "ionise it". This produces localized changes to the chemical properties of the material. If the radiation interacts with living material then the living material can be damaged. This damage is often repaired by the body but occasionally it is not leading to a change in the cell. It is changes in the body's cells that leads to the symptoms of radiation sickness and sometimes leads to cancer. These cancers can take years even decades to be noticeable.

89

https://www.gov.uk/government/uploads/system/uploads/attachment_data/file/467205/Basic_concepts_of_radiation_October_2015.pdf

90 https://en.wikipedia.org/wiki/Metric_prefix#List_of_SI_prefixes

91 http://hps.org/publicinformation/ate/faqs/radiationtypes.html

Radiation dose is a measure of the impact of radiation on the body. It is a complex concept because of the several different ways radiation can act on your body and the fact that different parts of the body have different levels of sensitivity.

To determine the level of harm caused by a radiation dose it is necessary to understand:

(1)   How much energy the radiation deposits in the affected tissue;

(2)   How much harm that deposited energy causes to those affected tissues;

(3)   The level of harm that is caused to the body by the harming of those affected tissues.

Taking these steps in turn:

(Step 1) **Absorbed dose (Gy)** is the energy (in Joules) absorbed in an amount of matter (in kg). Often used as mGy = 1/1000 Gy.

(Step 2) **Equivalent Dose (Sv)** is a measure of the ability to damage living tissue. Given by multiplying the Absorbed dose (Gy) by a radiation weighting factor (W_R) which accounts for different effectiveness of different radiations. Often used as mSv = 1/1,000 Sv ("milli-sievert") or µSv = 1/1,000,000 Sv ("micro-sievert").

(Step 3) **Effective dose (Sv)** is a measure of the ability harm an individual. Given by the Equivalent dose (Sv) weighted for the susceptibility and importance of the organ affected and summed over the whole body.

The unit used to measure a person's exposure to low levels of radiation is the sievert or milli-sievert (mSv) (1/1000 of a sievert) or micro-sievert (µSv) one millionth of a sievert.

The sievert is only used in radiological protection in the context of the risk of stochastic effects resulting from low exposure rates. For high dose rates, liable to cause deterministic effects, the grey (Gy) is used instead.

The average annual radiation exposure in the UK is about 2.7 mSv. This is the radiation dose we get from the materials around us, radiations from space, and from medical procedures (see table below).

Comparison of doses from sources of exposure[92]

Source of exposure	Dose
Dental X-ray	0.005 mSv
Chest X-ray	0.014 mSv
Transatlantic flight	0.08 mSv
Nuclear power station worker average annual occupational exposure (2010)	0.18 mSv
UK annual average radon dose	1.3 mSv
CT scan of the head	1.4 mSv
UK average annual radiation dose	2.7 mSv
US average annual radiation dose	6.2 mSv
CT scan of the chest	6.6 mSv
Annual exposure limit for nuclear industry employees	20 mSv
Level at which changes in the blood can be readily observed	100 mSv
Acute radiation effects including nausea and a reduction in white blood cells	1000 mSv
Dose of radiation which would kill about half of those receiving it within a month	5000 mSv

A short film on background ionising radiation can be found on YouTube[93]. A more detailed report can be found on the government website[94].

---

[92]https://www.gov.uk/government/publications/ionising-radiation-dose-comparisons/ionising-radiation-dose-comparisons
[93] https://www.youtube.com/watch?v=VCsJcfmL7TE&feature=player_embedded
[94] https://www.gov.uk/government/publications/ionising-radiation-exposure-of-the-uk-population-2010-review

B.2. Harm caused by radiation

There is no doubt that large doses of radiation over a short period of time are damaging to health. This has been known for a long time. What is less clear is the harm that might be done by small doses of radiation.

Two main types of radiation harm are known. The first type has a rapid impact while the second might take months to decades to be noticeable.

Short term effects (*deterministic effects*) are the effects that happen quickly during and after large exposures and include burning and other tissue damage. These have a threshold dose below which no effects are seen and above which the effect worsens with increasing dose. This is like sunburn which has no effect for short exposures but gets worse as you spend more time in the sun.

Deterministic effects include sickness, diarrhea, and anaemia which can be severe, even fatal, at high doses. Radiation doses of greater than a few grey (Gy) in a short time can make you very unwell and can kill.

It is unlikely that this class of effects will be noticed in any members of the public due to a nuclear emergency at a UK licensed site.

The second type of radiation effect is the longer term damage leading to effects such as cancer sometime after the exposure. This time delay can be as long as several decades after the exposure. In this case the probability of seeing the effect increases with dose received but the consequences are not affected by the dose (either you see the effect or you don't). The likelihood of these effects at low doses is very difficult to assess. The assumption currently used internationally is that the chances of an ill-effect such as a cancer decreases with decreasing dose and that there is no dose below which the chances are zero. This is the *linear no-threshold hypothesis* and is why scientists and government advisors are reluctant to say what a "safe" dose is. Effects of this type are called *stochastic effects*. Someone exposed to a low level of radiation may, or may not, suffer stochastic effects later in their life. It is impossible to say. It is also generally impossible to say that a cancer discovered later in life was due to the exposure or was fated to happen anyway.

Many people overestimate the danger of radioactivity and most would be surprised to hear that according to the Japanese Ministry of Health, Labour, and Welfare, 80% of the atomic bomb survivors who were younger than 10 years old in 1945 were still alive in 2015. Their mean age was then more than 80 years old[95]. So, even if you are unfortunate enough to be accidently exposed to radiation you still have a very good chance of a long and healthy life.

B.3. A brief history of Radiation Protection

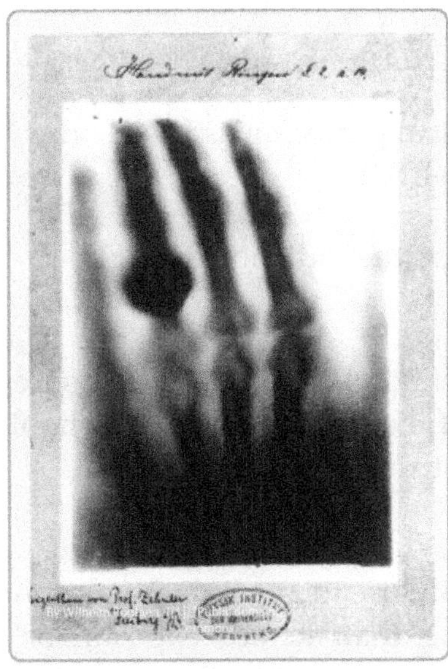

Ionising radiation was discovered in 1895 with the discovery of X-rays by Rontgen. The picture above is the first he published[96]. It shows his wife's hand. Almost immediately their value to the world of medicine was recognized and many people started to experiment with them. Glasgow Royal Infirmary had an X-ray department later that year[97].

It was not long before the harmful side effects became apparent. A headline in the New York Morning Journal of November 29th 1896 asked "Dose danger lurk in the X-rays?" and recounts the unpleasant experiences of Captain Webster after X-rays were used to locate a bullet in his body[98].

In the same year an American inventor Elihu Thomson deliberately exposed his finger to x-rays over several days and reported "pain, swelling, stiffness, erythema and blistering"[99]. In 1901 William Rollins reported that two Guinea Pigs exposed to X-rays for several hours a day died on the 8th and 11th day of the experiment[100]. It was clear that a balance had to be found in which the benefits of radiation outweighed the health costs.

The International Commission on Radiological Protection (ICRP) issued recommendations in 1928 which are updated periodically. Over the following years limits were placed on permissible levels of exposure, usually expressed as limits over short periods such as 1 day or 1 week, and designed to prevent deterministic effects (that is visible damage such as burning or reddening of skin).

---

[95] http://www.thelancet.com/pdfs/journals/lancet/PIIS0140-6736(15)00697-2.pdf
[96] https://www.nlm.nih.gov/dreamanatomy/da_g_Z-1.html
[97] http://www.auntminnieeurope.com/index.aspx?sec=ser&sub=def&pag=dis&ItemID=608942
[98] http://edison.rutgers.edu/NamesSearch/DocDetImage.php3
[99] http://www.physics.isu.edu/radinf/50yrs.htm
[100] http://radiationeffects.org/wp-content/uploads/2014/11/Kathren1996.pdf

In 1956 a meeting of the ICRP identified that these rules allowed large doses to be accumulated over time and proposed a quite complicated set of rules to limit longer term doses. These set age-dependent limits on the total accumulated dose of some key organs and limits to other organs over 13 consecutive weeks and over a year.

Since then the recommendations have evolved with the latest set of recommendations being published in 2007 (ICRP Publication 103).

B.4.   Radiation Protection today

Since zero risk is not achievable the current radiological protection is based on three main principles (from ICRP Publication 103 page 88):

The principle of justification: Any decision that alters the radiation exposure situation should do more good than harm.

The principle of optimisation of protection: the likelihood of incurring exposures, the number of people exposed, and the magnitude of their individual doses should all be kept as low as reasonably achievable, taking into account economic and societal factors.

The principle of application of dose limits: The total dose to any individual from regulated sources in planned exposure situations other than medical exposure of patients should not exceed the appropriate limits recommended by the commission.

The Commission now recognise three classes of radiation exposure:

- *Occupational exposure*, which is the exposure incurred at work, and principally as a result of work;
- *Medical exposure*, which is principally the exposure of persons as part of their diagnosis or treatment; and
- *Public exposure*, which comprises all other exposures.

ICRP 103 recommends that the source-related dose constraint for individuals exposed to a source that gives them little or no individual benefit but benefits society in general should be limited to 1 mSv. This provides the 1 mSv additional dose limit to members of the public living near to a nuclear site.
ICRP 103 also recommends that the highest planned residual dose in an emergency situation should be in the range of 20 mSv to 100 mSv. This gives the dose constraint of 100 mSv to a member of the public in the immediate stages of accident.

B.5.   What dose limits apply to me in an emergency?

The principles of radiological protection require that all doses to ionising radiation are kept as low as practical, taking economic and social considerations into account. Dose limitation is secondary to this intention.

If you live near a nuclear power station you can rest assured that all reasonably practical steps have been taken to reduce the radiation dose you receive as a result of the operation of the station. Discharges of radioactive are kept as low as reasonably practical with a dose limit of 1 millisievert (mSv) a year (UK Strategy for Radioactive Discharge, 2009)[101].

In a nuclear emergency doses higher than this limit are certainly possible. There are two very different stages of a nuclear emergency that require different ways of thinking about dose control. The first stage is the emergency stage where the radioactive plume is airborne and when people might be exposed to relatively high dose rates for a relatively short period of time and with little or no notice. The second stage is the recovery stage when the plume has gone but the environment is contaminated leading to dose rates that are above the pre-existing background rate, relatively low compared to the plume stage but expected to last for a longer period.

In the first stage it may be broadly acceptable to apply intrusive countermeasures such as asking people to shelter, to evacuate or to take stable iodine tablets. In the second stage the aim is to achieve a "new normality" where life can continue without excessively onerous constraints.

Current advice from the ICRP (ICRP Publication 103) defines emergency exposure situations as "unexpected situations that may require urgent protective actions, and perhaps also longer-term protective actions, to be implemented". The advice states that emergency plans should aim to keep residual doses at or below 100 mSv. Residual doses being the additional dose to individuals resulting from the emergency after countermeasures have been applied. But remember that the dose response is considered to be linear. A dose of 99 mSv is not entirely "safe" and a dose of 101 mSv is not significantly more "unsafe" instead each unit of dose carries a risk of long term effect. The ICRP states that consideration should be given to reducing any projected doses in the range 20 mSv to 100 mSv with increasing efforts made to prevent doses at the higher end of the range.

---

[101]

https://www.gov.uk/government/uploads/system/uploads/attachment_data/file/249884/uk_str ategy_for_radioactive_discharges.pdf

After the plume has gone the concern switches to the contamination in the environment which may continue to give a radiation dose to people in the area. ICRP defines this as *an existing exposure situation,* one that already exists when a decision on control has to be taken. Radon in dwellings and offices is an example of an existing exposure that is managed in the UK[102]. For existing exposure situations the ICRP recommends reference levels of typically not greater than 20 mSv and suggest that interventions might be appropriate to reduce dose-rates in the range 1 - 20 mSv/year.

After the Fukushima event the public found it hard to accept that the dose limit applied to them changed from 1 mSv per year before the event, to 100 mSv during the event and its immediate aftermath and then to 20 mSv per year after the event but this was confirmed by the ICRP in a letter[103] dated March 21, 2011 and does make sense if you remember that these are not "safe" and "unsafe" levels and that the intention is to keep doses as low as practical given the situation and without unnecessarily ruining lives.

The IAEA document communicating with the public in a nuclear or radiological emergency[104] has this to say about the 100 mSv dose level.

> *"At doses below 100 mSv there would not be any detectable cancers or other severe health effects even to the foetus. The termination of a pregnancy at foetal doses of less than 100 mSv is NOT justified based upon the radiation risk. An increase in the cancer rate has not been detected in any group of people who received a whole body dose from external exposure below about 100 mSv"*

According to the World Health Organisation (WHO) *"Everybody has a chance of having a cancer (incidence) and/or dying from cancer (mortality) over the course of her/his lifetime. This is the so-called "lifetime baseline risk" (LBR). The additional risk of premature incidence or mortality from a cancer attributable to radiation exposure is called the "lifetime attributable risk" (LAR). The LAR is an age- and sex-dependent risk quantity calculated by using risk models derived from epidemiological studies."*[105]

---

[102] http://www.ukradon.org/
[103] http://www.icrp.org/docs/fukushima%20nuclear%20power%20plant%20accident.pdf
[104] http://www-pub.iaea.org/MTCD/publications/PDF/EPR-Communcation_web.pdf
[105] http://www.who.int/ionizing_radiation/pub_meet/radiation-risks-paediatric-imaging/en/

It is important to remember that cancer is all too common. According to Cancer Research "*1 in 2 people born after 1960 in the UK will be diagnosed with some form of cancer during their lifetime*"[106]. That is, the lifetime baseline risk (LBR) is about 0.5 (or 1 in 2). Different cancers have different incident rates and different impacts. For example BEIR shows breast cancer affecting 12,000 women in 100,000, killing 3,000 of them costing those killed an average of 15 years of life whereas thyroid cancer affects 230 males and 550 females per 100,000, killing 40 males and 60 females and costing those killed an average of 12 years of life[107].

The same BEIR publication estimates that if 100,000 people of mixed ages were to be exposed to 0.1Gy (or 100 mSv) there would be an additional 310 cases of breast cancer (95% confidence limits of 160 - 610) compared to the 12,000 that would be expected without the exposure. It would be impossible to know if a particular cancer was associated with the exposure or not and very difficult to prove an excess of cancers at these levels. It can be concluded that an exposure of 0.1 Gy makes no practical difference to a women's chance of getting breast cancer (if 100,000 women were exposed to 0.1 Gy (or 100 mSv) the expected number of breast cancers would rise from about 12,000 to about 12,310. Given the statistical variations in populations it would very difficult to confirm an excess of breast cancers).

For thyroid cancer the estimated additional number of cases is 21 for males and 100 for females compared to the background of 230 and 550 respectively. So this level of radiation would make a small difference to the chance of getting thyroid cancer but it would require a large population to be sure the effect was provable.

---

[106] http://www.cancerresearchuk.org/health-professional/cancer-statistics/risk/lifetime-risk
[107] https://www.nap.edu/read/11340/chapter/14

Appendix C - Other types of event

Nuclear accidents are not the only situations in which the public can be exposed to additional doses of ionising radiation. This section briefly looks at the consequences of:

- Dirty Bombs - devices left by terrorists which use convention explosives to spread a radioactive material;

- Orphan source - where a radioactive source escapes control and exposes people as they go about their daily life;

- Nuclear satellite re-entry;

- Atomic bomb.

In all these cases protection of the public relies on the controls of time, distance and shielding in the immediate event and decontamination as the event progresses.

C.1   Dirty Bomb

A Dirty Bomb or radiological dispersal device (RDD) is a device that combines conventional explosives with radioactive material. It is not a nuclear explosion. The intention might be to spread alarm, based on the well-known fear of radiation, and to make recovery from the conventional bomb more difficult i.e. to deny access to an area for a prolonged period causing economic and social disruption.

To detonate a dirty bomb in a public space the terrorists would need the technology to deliver and initiate a conventional bomb (all too often groups and individuals succeed in this endeavour) and they would also need to obtain a radioactive source without coming to the attention of the authorities. There are many radioactive sources in use in hospitals and in industry but they are subject to tight control. They also tend to be in non-dispersible form; as metals or ceramics or tightly encapsulated. The stronger sources would be powerful enough to make anyone tampering with them or carrying them outside their heavy shielding ill quite quickly. The terrorist group would have to find a way to handle the material and to make it dispersible in order to make a successful dispersal device as opposed to a scary news story.

The immediate danger would be the bomb blast and priority should be given to treating those with serious injuries and providing support to the other injured people and worried bystanders. At this stage no one would know the bomb contained radioactive material unless tipped off by the terrorists. These days first responders carry radiation detectors so any significant contamination should be discovered quite quickly.

Persons contaminated by a dirty bomb are unlikely to be a health hazard to those treating them. A possible exception to this could be if the victims have highly radioactive fragments embedded in their wounds. These can be readily detected with hand held radiation detectors and should be removed with long tweezers, avoiding direct contact, and placed in lead lined boxes or at least stored a distance from people if shielded boxes are not available.

The stabilisation of those with life threatening injuries takes priority over decontamination. For other casualties the bulk of any contamination is likely to be removed by removing their outer clothes and bare skin can be decontaminated by brushing, showering or washing.

Advice to individuals affected by a dirty bomb can be found in a FEMA factsheet[108].

If you are unfortunate enough to be near the scene of a bomb explosion in a city street and there is nothing you can do to help the victims then the best advice is to leave the immediate area and, in particular, avoid breathing in any dust produced by the explosion - run upwind or across wind rather than downwind. If there is an obvious assembly point for those near the scene at the time then go there.

If you have dust from the explosion on your clothes remove your outer layers of clothes. Wipe the dust from your skin as best you can without damaging the skin or making it red. If possible wash your hair trying not to get the runoff into your mouth, nose, eyes or ears. If it was a dirty bomb then the authorities will provide suitable advice, monitoring and decontamination as quickly as they can organise it but the sooner you can do the basics yourself the better. See decontamination infographic[109] for sensible advice on decontamination.

If you are in an undamaged building after the blast then it may be best to stay there with the doors and windows shut, air conditioning off until advised by the authorities to move.

In all events you should listen to local radio and television for news and official advice. Use social media and phones to contact family and friends but think carefully before believing everything you hear or read on these systems.

There will almost certainly be a helpline advertised to phone for advice. The usual NHS helplines (NHS 111) will be available in the UK.

---

[108]https://www.fema.gov/media-library-data/20130726-1621-20490-3999/dirtybombfactsheet_final.pdf
[109] https://emergency.cdc.gov/radiation/pdf/Infographic_Decontamination.pdf

C.2   Orphan source

An orphan source is one that has escaped control. These can be very powerful sources, although the term does also apply to weak sources. There have been a number of cases where sources have been lost and gone on to cause significant harm. The IAEA have recorded 68 such incidents[110].

The worst to date was the accident in Goiania in 1985 which resulted from the theft of a 50.9 TBq caesium-137 source from a partly demolished private medical facility[111]. This was dismantled by people unaware of the risks. The glittering powder attracted interest and several people used on their skin as decoration. A few weeks later, after a doctor recognised the effects of radiation in patients presenting with burns and sickness, the area effected was monitored and evacuated. People sheltered at a sports facility for several days awaiting monitoring. About 110,000 persons were monitored. Of these, 249 were shown to be contaminated, 150 of which required further medical care. 14 required intensive care, 4 died. The decontamination of the area took six months, some flats were evacuated pending decontamination and some demolished. Forty five public places (including streets, squares and shops) were decontaminated.

Another example of an orphan source emergency, this time a by-product of an assassination, was the Polonium-210 incident in London, 2006. Mr Litvinenko was admitted to hospital on 3rd November with vomiting, diarrhoea and abdominal pain. After some delay it was determined that he had been poisoned with about 1 GBq of ^{210}Po, a pure alpha emitter that is not detected by many types of radiation detector in common use. The identification of exposed individuals and identification and decontamination of contaminated places was a major undertaking[110].

If an intact source were to be left in a public area then people near that source would be exposed. The effects of that exposure would depend on the strength of the source, how near people got to it and how long they stayed near it. This is illustrated by Table 3 which shows the doses received at different times and distances from an unshielded 50 TBq Cs-137 source (Gamma ray dose constant of 1.02E-04 mSv/hr^{-1} per MBq at 1m)[112]. The exposure time on the vertical axis range from 5 minutes to 2 hours and the distance on the horizontal axis from 1 to 10 metres. Doses in the top left hand corner (long exposure and close to) are over 5 Sv and could well be fatal. Doses in the bottom right corner are below 100 mSv and people is this range would probably not show any symptoms. The two middle blocks correspond to 1 - 4 Sv (likely to be ill, could be fatal at high end) and 100 mSv to 1 Sv (could show some symptoms but likely to recover with medical care).

---

[110]   http://www-pub.iaea.org/books/IAEABooks/8920/Lessons-Learned-from-the-Response-to-Radiation-Emergencies-1945-2010

[111] http://www-pub.iaea.org/mtcd/publications/pdf/pub815_web.pdf

[112] https://www.orau.org/documents/ivhp/health-physics/ornl-rsic-45.pdf

Dose in Sv staying at distance on x-axis for time on y-axis

Staytime in minutes	1	2	3	4	5	6	7	8	9	10
120	10.170	2.543	1.130	.636	.407	.283	.208	.159	.126	.102
115	9.746	2.437	1.083	.609	.390	.271	.199	.152	.120	.097
110	9.323	2.331	1.036	.583	.373	.259	.190	.146	.115	.093
105	8.899	2.225	.989	.556	.356	.247	.182	.139	.110	.089
100	8.475	2.119	.942	.530	.339	.235	.173	.132	.105	.085
95	8.051	2.013	.895	.503	.322	.224	.164	.126	.099	.081
90	7.628	1.907	.848	.477	.305	.212	.156	.119	.094	.076
85	7.204	1.801	.800	.450	.288	.200	.147	.113	.089	.072
80	6.780	1.695	.753	.424	.271	.188	.138	.106	.084	.068
75	6.356	1.589	.706	.397	.254	.177	.130	.099	.078	.064
70	5.933	1.483	.659	.371	.237	.165	.121	.093	.073	.059
65	5.509	1.377	.612	.344	.220	.153	.112	.086	.068	.055
60	5.085	1.271	.565	.318	.203	.141	.104	.079	.063	.051
55	4.661	1.165	.518	.291	.186	.129	.095	.073	.058	.047
50	4.238	1.059	.471	.265	.170	.118	.086	.066	.052	.042
45	3.814	.953	.424	.238	.153	.106	.078	.060	.047	.038
40	3.390	.848	.377	.212	.136	.094	.069	.053	.042	.034
35	2.966	.742	.330	.185	.119	.082	.061	.046	.037	.030
30	2.543	.636	.283	.159	.102	.071	.052	.040	.031	.025
25	2.119	.530	.235	.132	.085	.059	.043	.033	.026	.021
20	1.695	.424	.188	.106	.068	.047	.035	.026	.021	.017
15	1.271	.318	.141	.079	.051	.035	.026	.020	.016	.013
10	.848	.212	.094	.053	.034	.024	.017	.013	.010	.008
5	.424	.106	.047	.026	.017	.012	.009	.007	.005	.004

Distance from source in metres

Table 3 Variation of dose with time and distance

This sort of table can be used by the emergency services and other responders to estimate how long they can stay near the source for a particular dose limit although it is better to use measurements of the dose rates rather than estimates. Their stay-time could be increased if they could put some shielding between them and the source.

The first signs of a powerful source being left in a public area might be people presenting at hospitals, doctors surgeries or calling for an ambulance with symptoms of acute radiation syndrome or unexplained burns. Importantly, if the source was intact then no one would be contaminated and the casualties would pose no hazard to those helping them or to any medical facility they were taken to.

The best defence from orphan sources is time, distance and shielding. Recognise the radiation warning signs (Figure 7) and if in doubt stay away from things that you think might be radioactive.

Figure 7 Radioactive material signs

In the UK the police have access to the NAIR plan (National Arrangements for Incidents Involving Radioactivity)[113]. This provides quick and widely available assistance to the police and other emergency services where no radiation expert is otherwise available. Assistance is drawn from hospitals, the nuclear industry and government departments.

C.3   Nuclear satellite re-entry

It is claimed that between 1967 and 1988 the Soviet Union launched over 30 fission reactors into space while the USA launched one to provide power for satellites[114]. In addition to reactors there have been a number of radioisotope thermoelectric generators (RTGs) launched. These use the heat generated by the decay of radionuclides (generally Pu-238 or Am-241) to provide power. Many of these systems are still up there. There are supposedly "parked" in high orbits and unlikely to crash to earth for several hundred years so a radioactive satellite re-entry is inevitable at some time in the future but apparently unlikely in the next few decades at least. By the time they are expected to crash there will be less radioactivity (Pu-238 has a half-life of 87.7 years, Am-241 432.2 years).

Reassuringly IAEA Safety Series 119[115] (superseded) states that launched systems are generally designed to fragment and burn-up on re-entry and that *"If the total inventory of a 100 kW(th) reactor is deposited homogeneously over an area of 100 000 km 2, the net average external dose rate increase in the central part of the footprint would be about 0.05 µSv/h, significantly less than the natural background."*

The experience of the Cosmos 954 re-entry in January 1978 over the Northwest Territories of Canada gives an idea of the issues that might arise. It spread debris over 124,000 square kilometres. Some of this debris was sufficiently radioactive to be hazardous to health.

---

[113] https://www.gov.uk/guidance/national-arrangements-for-incidents-involving-radioactivity-nair
[114] http://www.world-nuclear.org/information-library/non-power-nuclear-applications/transport/nuclear-reactors-for-space.aspx
[115] https://gnssn.iaea.org/Superseded Safety Standards/Safety_Series_119_1996.pdf

The Canadian Nuclear Safety Commission reports[116] that *"by October 1978, over 4,000 flakes of radioactive debris were recovered from the region, including core fragments. It took nine months, 4,500 hours of flying time and $13,970,143 to recover from the crash of the COSMOS 954"*.

We can conclude that a nuclear satellite crashing on the UK would be a wide area, long term emergency, would need some clear communication of the risks to the public and a major radionuclide survey of the affected area, and probably beyond.

As a member of the public you should listen to the advice being given by the authorities which would probably be to carry on life as usual but stay away from and report any visible debris that you might see.

C.4   Nuclear bomb

A nuclear bomb detonation in the UK is considered to be a very low probability. The 2017 National Risk Register[117] mentions the use of an improvised nuclear device by terrorists *"resulting in much greater numbers of casualties and widespread, long-term impacts of a magnitude above all other terrorist attacks"* but the CBRN response discussed on the Centre for Protection of National Infrastructure website[118] is basic.

The consequences of such as detonation would depend on the nature of the attack. A single improvised bomb is likely to have a relatively low explosive yield, a nuclear device stolen from one of the nuclear powers would pose a far greater threat and an all-out attack from a superpower would be another matter entirely.

The explosive power of a nuclear bomb is measured in terms of the amount of TNT that would be required to give a similar size explosion. (A ton of TNT releases about 4.184×10^9 joules) There are a wide range of explosive yields available from nuclear devices (see list below adapted from Wikipedia[119]).

Name	Approx yield*	Details
David Crocket	0.01 kT	US tactical nuclear weapon
Hiroshima (Little boy)	13 - 18 kT	1st atomic bomb attack - 6/8/1945

---

[116]   http://nuclearsafety.gc.ca/eng/resources/canadas-nuclear-history/past-presidents/alan-prince.cfm?pedisable=true

[117]
https://www.gov.uk/government/uploads/system/uploads/attachment_data/file/644968/UK_National_Risk_Register_2017.pdf

[118] https://www.cpni.gov.uk/chemical-biological-radiological-and-nuclear-cbrn-threats-0

[119] https://en.wikipedia.org/wiki/Nuclear_weapon_yield

Nagasaki (Fat man)	18 - 23 kT	2nd atomic bomb attack - 9/8/1945
Hurricane	25 kT	1st British atomic explosion - 3/10/1952
Castle Bravo	15,000 kT	Largest nuclear detonation by USA 1/3/1954
Tsar Bomba	50,000 kT	Largest ever nuclear explosion (USSR) 31/10/61
Test No. 6	3,300 kT	China 17/6/67
Canopus	2,600 kT	France 24/8/1968
Pokhran-II	45 - 50	India 11/5/1998
Chagai-I	40	Pakistan 28/5/1998
2017 North Korean nuclear test	200 - 300	North Korea 3/9/2017

*   Some of these values must be rather speculative.

A nuclear explosion has a number of identifiable steps:

- Fireball - caused by the release of energy;

- A wave of thermal radiation - very fast moving and capable of blinding, burning or killing outright. First degree burns (similar to bad sun-burn) may be expected out to 3.2 km from a 10 kT bomb; second degree burns (blistering, possible shock and death if extensive and untreated) out to 2.4 km and third degree (destruction of tissue, shock and death without specialist treatment) out to 1 km (FEMA[120]);

- A shock-wave that can cause considerable physical damage to anything in its way. At 0.4 km from a 10 kT explosion the wind could be 160 miles an hour, sufficient to suck people out of office blocks;

- An electromagnetic pulse - this can damage electricity supply and telephone systems as well as computers and other electrical systems. Wireless communications may be disrupted for a while even if the hardware is unaffected. A blast high in the atmosphere could damage any satellites within range. This is considered to be a tactical use of nuclear weapons, and one to be concerned about as it greatly damages the ability of a modern society to function without causing so much longer term fallout;

- A mushroom cloud as everything within the fireball is vapourised and carried upward. This can include great quantities of dust if the blast is low enough to hit the ground and produce a crater;

---

[120] https://emilms.fema.gov/IS3/FEMA_IS/is03/REM0502100.htm

- Fallout - caused by the dusts and vapours produced cooling, condensing and returning to ground level. This fallout can be intensely radioactive although the levels of radioactivity will be dropping rapidly, at least at first;

To survive a nuclear bomb it is necessary to survive a number of deadly effects:

- The air blast (shock wave) caused by the energy released locally heating up the air and causing it to expand rapidly;

- The thermal radiation produced in the explosions and the interactions with the atmosphere;

- The neutron radiation produced in the explosions;

- The gamma radiation produced in the explosions;

- Any fire storm resulting from the blast;

- The beta/gamma radiation from decay of fission products and neutron induced radioactivity within the fallout.

The initial pulse of radiation may be sufficient to kill, or at least blind, people beyond the zone that is devastated by the blast. Survival is dependent on being far enough away from the bomb or well shielded. Those looking towards the bomb are obviously more likely to have their eye sight damaged, although this may be temporary for those "lucky" enough to be further away.

The effects of the blast travel more slowly than the flash, can destroy buildings and kill people over an extensive range. Again survival is about distance and sheltering. Shelters would have to be robust to survive the blast. Underground would be preferred.

The blast can cause changes in air pressure, both sudden increases and decreases, which can be sufficient to kill. The blast and resulting fire can deplete the oxygen supply leading to death.

For the unprepared and unwarned survival of the blast is a matter of luck - not being too close to the bomb when it goes off or fortuitously being shielded. Thereafter survival of the next few minutes could be enhanced by quickly (very quickly!) finding better shelter from the thermal and radiation blast and their effects. If you can't find shelter within a few seconds then lie on the ground and cover your head if possible. Do not look at the flash or fireball if you can help it.

The next issue is fallout - the radioactive dust and gases produced by the explosion. Survival is about putting as much distance or shielding between individuals and the fallout. Evacuations should be away from the fallout. Any survivors in shelter will need advice, reassurance, and probably, medical attention. A good source of such advice is the USA Ready Government web-site[121].

There are two types of shelter associated with surviving a nuclear blast. A **Blast Shelter** is designed to protect the occupants from the initial blast, thermal energy and radiation. A **Fallout Shelter** is designed to protect the occupants from the radiation in the fallout. One suitably built shelter can perform both functions. The UK Government's rather old "protect and survive" booklet[122] gives some advice on how to position and build an improvised shelter within your own home if you had reason to believe that an attack could take place in the next few days. The document states that *"Your best protection is to make a fall-out room and build an inner refuge within it"*. The inner refuge is an attempt to surround yourself with material that will absorb the high levels of radiation present in the first few hours after detonation, you are recommended to stay in it for a long as possible and at least two days. Sand, earth and books are recommended. The advice is to take in enough drinking water for everyone to have two pints a day for fourteen days and the same again for hygiene purposes. Food supplies for fourteen days are also recommended. More hints on the building of shelters are given in Domestic Nuclear Shelters[123], also by the UK Government and also rather dated.

For the few days or weeks after the explosion avoidance of the fallout is necessary to prevent the exposure to a large radiation dose. Evacuation away from the dust falling from the sky followed by decontamination is the quick route out. Staying in a fallout shelter until the radiation levels have dropped low enough not to be a great hazard is the slow way out. Those exposed to high doses of radiation or injured in the blast may be unlikely to survive without prompt medical care.

FEMA report a 7-10 rule[124]: *The 7:10 Rule of Thumb states that for every 7-fold increase in time after detonation, there is a 10-fold decrease in the exposure rate. In other words, when the amount of time is multiplied by 7, the exposure rate is divided by 10. For example, let's say that 2 hours after detonation the exposure rate is 400 R/hr. After 14 hours, the exposure rate will be 1/10 as much, or 40 R/hr.*

[121] https://www.ready.gov/nuclear-blast
[122] http://www.atomica.co.uk/main.htm
[123] http://www.atomica.co.uk/shelters/main.htm
[124] https://emilms.fema.gov/IS3/FEMA_IS/is03/REM0504050.htm

Recently published advice to potential responders to a nuclear detonation in the USA can be found on the internet[125], a fuller health and safety guide is also available[126].

Yield (kT)	Shockwave (km)	Heat (km)	Initial Radiation (km)	Fallout radiation (downwind) (km)
1	0.3	0.6	0.8	Up to 6
10	0.6	1.8	1.3	Up to 10

Significant effects are 50% mortality from shockwave and heat and a radiation dose of 4 Sv

Table 4 - Range of significant effects from atomic bomb (km)

From[127]

C.5   Nuclear risks in context

This document has focused on nuclear risks and, within that, the risks of an accident at a nuclear power station. But these sites are designed and operated to keep risks at a low level. There are many other risks facing us in day to day life.

The UK government maintains a watching brief of the risks posed to society and uses this to determine where national resources should be used to minimise risks and prepare for emergencies. It also requires that local authorities undergo a similar process for their area.

The *National Risk Register of Civil Emergencies* is now an annual publication that is available on line (the 2017 version is the latest as this is being written[128]). The local documents are prepared by the Local Resilience Forums[129].

---

[125] https://www.dhs.gov/sites/default/files/publications/Quick%20Reference%20Guide%20Final.pdf

[126] https://www.dhs.gov/sites/default/files/publications/IND%20Health%20Safety%20Planners%20Guide%20Final.pdf

[127] https://www.dhs.gov/sites/default/files/publications/prep_nuclear_fact_sheet.pdf

[128] https://www.gov.uk/government/uploads/system/uploads/attachment_data/file/644968/UK_National_Risk_Register_2017.pdf

[129] https://www.gov.uk/guidance/local-resilience-forums-contact-details

The National Risk Register shows two matrices summarising natural and malicious threats (Figure 8 and Figure 9). This shows the estimated likelihood of occurrence along the bottom against the impact along the side. The first matrix shows pandemic flu possibly having the worst impact of any event considered and also being relatively likely. Because of the perceived risks of pandemic flu a lot of planning has taken place to ensure that the National Health Service is as prepared as possible.

Wide spread electrical failure and river and coastal flooding and cold and snow are all seen has having potentially damaging consequences, with cold and snow being more likely as these.

The potential impact of heat wave should be noticed. In August 2003 a UK heatwave lasted 10 days and resulted in over 2,000 deaths. Temperatures reached what was then a record 38.5°C in Faversham, England and 33°C in Anglesey, Wales. High temperature records are now being broken with increasing frequency. UK heatwave planning can be found on the Government web-site[130]. This includes a check-list for people to use at home[131]. See also NHS choices[132].

A relative newcomer to the national risk register is Space Weather[133]. This relates to a number of phenomena originating in the sun or space that pose risks to technology, infrastructure, and electrical systems. That is, they can damage satellites, including GPS and communications, and the power grids. In extreme cases areas may be without electrical power for long periods of time. In 1989 a space storm disrupted electricity networks in Quebec in Canada, which led to a 9 hour power cut across the province. There are fears that repairing the national Grid after such a space storm could take several months.

On nuclear risks the risk assessment document states that *"Nuclear sites are designed, built and operated so that the chance of accidental releases of radiological material in the UK is extremely low. Historical accidents include Windscale (UK) in 1957, Three Mile Island (US) in 1979, Chernobyl (Ukraine) in 1986 and Fukushima (Japan) in 2011. Of these, Chernobyl and Fukushima were the most severe"*. In the matrix, nuclear accidents are included with "industrial and urban accidents". This set of events is potentially as damaging as surface water floods, infectious diseases, space weather, poor air quality and heat wave but less likely.

---

[130] https://www.gov.uk/government/publications/heatwave-plan-for-england
131
https://www.gov.uk/government/uploads/system/uploads/attachment_data/file/525361/Beatth eheatkeepcoolathomechecklist.pdf
[132] https://www.nhs.uk/livewell/summerhealth/pages/heatwave.aspx
[133] https://www.gov.uk/guidance/space-weather-and-radiation

In malicious events chemical, biological, radiological and nuclear attacks are seen as the events of highest consequence but are considerable less likely than many other forms of attack.

For my local area in Gloucestershire the Local Risk Register[134] has pandemic disease, poor air quality caused by volcanic activity in Europe, and total failure of the UK's National Grid (electricity network) as the leading risks. Other significant risks include various kinds of industrial accidents, severe weather events, human and animal health issues, industrial action and public disorder.

Thus we, as a nation, as local communities and as families and individuals should be prepared to cope with a range of accidents and malicious activity because if we are prepared we are more likely to survive and more likely to get back to normal life more quickly. Fortunately many of the sensible steps to take are common to several types of emergency.

The American equivalent of the National Risk Register seems to be *the Strategic National Risk Assessment*. A summary of this is available on the internet[135].

---

[134]   http://glosprepared.co.uk/wp-content/uploads/2015/12/Gloucestershire-LRF-Community-Risk-Register-2014-15.pdf
[135]https://www.dhs.gov/xlibrary/assets/rma-strategic-national-risk-assessment-ppd8.pdf

<table>
<tr><td rowspan="7">Overall relative impact score</td><td>5</td><td></td><td></td><td></td><td>Pandemic flu</td><td></td></tr>
<tr><td>4</td><td></td><td></td><td>Widespread electrical failure. Coastal and river flooding</td><td>Cold and Snow</td><td></td></tr>
<tr><td>3</td><td></td><td>Transport Accident. Industrial and urban accident.</td><td>Surface water flooding</td><td>Infectious disease. Space weather. Poor air quality. Heatwave</td><td></td></tr>
<tr><td>2</td><td></td><td>Wild fire.</td><td>Animal disease. Drought. Industrial action.</td><td>Volcano. System failure. Public disorder. Storms & gales.</td><td></td></tr>
<tr><td>1</td><td></td><td>Earthquake.</td><td></td><td></td><td></td></tr>
<tr><td></td><td>Between 1 in 20,000 and 1 in 2,000</td><td>Between 1 in 2,000 and 1 in 200</td><td>Between 1 in 200 and 1 in 20</td><td>Between 1 in 20 and 1 in 2</td><td>Greater than 1 in 2</td></tr>
<tr><td colspan="6" style="text-align:center">Likelihood of occurring in next five years</td></tr>
</table>

Figure 8 Risk Matrix based on National Risk Register.

Overall relative impact score	Between 1 in 20,000 and 1 in 2,000	Between 1 in 2,000 and 1 in 200	Between 1 in 200 and 1 in 20	Between 1 in 20 and 1 in 2	Greater than 1 in 2
5		Large scale CNBR attack			
4					
3		Attack on infrastructure	Cyber-attack on infrastructure. Small scale CBRN		Attach on crowed space or infrastructure.
2				Cyber-attack on services	
1					
			Likelihood of occurring in next five years		

Figure 9 - Risk Matrix based on National Risk Register

Appendix D - Key Radionuclides

This section gives an outline of the properties of some of the key radionuclides you might hear about in the event of a nuclear accident.

D.1. Iodine

Iodine is a chemical element with symbol I and atomic number 53. The heaviest of the stable halogens, it exists as a lustrous, purple-black metallic solid at standard conditions that sublimes readily to form a violet gas (Wikipedia).

It is produced in nuclear reactors as a fission product.

For example I-135 is produced in about 2.9% of fissions (IAEA WIMS Library[136]). It is also produced as a result of the decay of Te-135 ($t_{1/2}$ 19s) taking its effective yield up to over 6%. This is a very high yield compared to most fission products. Because of its short half-life of 6.75 h it will reach an equilibrium level in a reactor at power where its rate of production equals its rate of decay.

Iodine is relatively volatile (Melting point: 113.7 °C, Boiling point: 184.3 °C) so is more mobile than many chemicals in nuclear reactors and, as a result, has a relatively high release fraction from damaged fuel and tends not to deposit on the reactor internals like some less volatile chemicals.

Many chemical forms of iodine can pass through the physical filters designed to reduce the release of particulate radioactivity in normal operation and in accident conditions. Some systems have iodine filters composed of volumes of activated charcoal (i.e. charcoal impregnated with chemicals that react with iodine). If these can be used then the iodine release can be greatly reduced.

Iodine is an essential nutrient used to make thyroid hormones which control metabolism among other functions. Because of its role in metabolism the body is very efficient at retaining any iodine either eaten or breathed in and collects it in the thyroid gland in the throat. Once in the thyroid the radioiodine gives a continuing radiation dose to this organ and the surrounding tissues.
The thyroid gland is known to be radio-sensitive in children and babies, less so in people over about 40 years old.

Given the information above it is easy to see why iodine is a problem in nuclear accidents. It is efficiently produced in the reactor, it is volatile allowing a high release fraction in accidents, a high fraction of it can pass through particulate filters, the body holds on to it and then concentrates it in a radio-sensitive organ.

---

[136] https://www-nds.iaea.org/wimsd/fpyield.htm

Fortunately we can significantly reduce the health impact of radio-iodine by taking a large dose of stable iodine and saturating the thyroid gland causing any subsequent intake to be excreted.

With the short half-life of most of the iodine isotopes it is not a long term problem. There is very little of it left after about a month after a reactor has shutdown which is why stable iodine is not used as a countermeasure around decommissioning sites. In a reactor accident there is a great deal of concern about iodine in the short term. This concern includes trying to minimise the inhalation of iodine during the release (shelter and evacuation), blocking inhalation that cannot be prevented (stable iodine) and trying to minimise ingestion in the days and weeks after the release (controls on some food notably milk). Stable iodine blocking can be used to reduce the risk of taking contaminated milk but it is preferred to prevent uptake.

Key Isotopes (half-life) I-131 (8 days), I-132 (2.3 hours), I-133 (20.8 hours), I-134 (52.5 minutes), I-135 (6.52 hours).

D.2. Caesium

Caesium is a chemical element with symbol Cs and atomic number 55. It is a soft, silvery-gold alkali metal with a melting point of 28.5 °C (83.3 °F), which makes it one of only five elemental metals that are liquid at or near room temperature (Wikipedia).

It is a fission product with cumulative fission product yields of Cs-134 7.6560E-08%, Cs-135 4.9265E-06%, Cs-137 6.3429E-02%. Because of the fixed ratio in their production and the different half-lives the ratio of Cs-137 to Cs-134 in a sample can be used to determine when it was produced.

Caesium is relatively volatile and has a relatively high release fraction from reactors in accident conditions.

Both Cs-134 and Cs-137 make a significant contribution to off-site doses in a reactor accident. This is both in the short term via external radiation from the plume and inhalation and in the longer term via food and ground contamination.

Key isotope (half live): Cs-134 (2.06 years), Cs-137 (30 years).

D.3. Americium

Americium is a synthetic chemical element with symbol Am and atomic number 95. It is a transuranic member of the actinide series. It does not appear in nature but is produced in nuclear reactors, not as a fission product but as the result of a series of neutron captures and decays within the fuel.

Americium can be a significant contributor to off-site dose in the event of a nuclear reactor accident. Am-241 which has a half-life of 432 years is the significant Americium isotope. Being an alpha emitter, this is not an external hazard but is particularly dangerous if ingested or inhaled.

Am-241 demonstrate the dangers of using partial information to assess danger. It is not detected by the majority of dose meters including those commonly used in perimeter monitoring systems, hand held detectors (other than specialized alpha detectors) and personal dosimeters yet can give significant inhalation or ingestion doses following a reactor accident.

Key isotopes (half-life): Am-242 (16 hours), Am-244 (10 hours).

D.4. Noble Gases

There are a number of radioactive isotopes of krypton and xenon that are fission products. These are very unreactive gases. Because of their chemical and physical properties these gases can pass through the barriers created to retain fission products such as fuel matrix, fuel clad, primary circuit and filters with relative ease. As a result their "release fraction" is very much higher than for other fission products.

Because they are inert gases they are not retained in the lungs. Any that you breathe in you immediately breathe out. Thus they give no inhalation dose. They do give an external radiation hazard in the form of cloud gamma as the plume passes but no persistent ground gamma dose because they do not settle on surfaces and are not incorporated into living tissue to any great extent.

Noble gases are often the first radioisotopes to escape from reactor sites. In the Three Mile Island accident almost all the activity released was in this form. Unlike Americium, which is hard to detect but gives high doses, the noble gases are easy to detect with dose-rate meters but gives very little inhalation dose. This is why dose-rate meters which measure external radiation do not give a good indication of the dose rate to people within a plume and subject to an inhalation dose.

Key isotopes (half live): Kr-85 (10.7 years), Kr-85m (4.5 hours), Kr-87 (76 minutes), Xe-133 (5.2 days),   Xe-135 (9 hours), Xe-138 (14 minutes).

D.5. Plutonium

Plutonium is an element that is produced in nuclear reactors when uranium nuclei absorb neutrons without fissioning. It is radioactive. Its symbol is Pu and it has atomic number 94. It is an actinide metal of silvery-grey appearance that tarnishes when exposed to air, and forms a dull coating when oxidized.

Plutonium is not very volatile and tends to have a very low release fraction in nuclear accidents. However, isotopes of plutonium tend to be alpha emitters and have high dose per unit intake factors for both inhalation and ingestion. Long half-lives means that any plutonium released would be persistent in the environment.

Key isotopes (half live): Pu-238 (87.4 years), Pu-239 (24,000 years), Pu-240 (6,500 years), Pu-241 (14 years), and Pu-242 (3.7×10^7 years).

Appendix E Further reading

This section suggests further reading to support the text.

E.1   How to survive a nuclear war

Protect and Survive[137]. This is a leaflet that was published in 1980 with the intention of distributing it to every household in the UK if we ever faced the imminent threat of a nuclear war. It provides advice on what a nuclear weapon detonation would look like and how to increase your chances of surviving the blast and the effects of fallout. I hope it is never distributed.

Domestic shelters providing protection against nuclear explosions[138]. This is a companion document to the one above giving more detailed advice on how to build yourself a bomb shelter.

The US government take on this subject can also be found on the internet[139],[140].

E.2   Introduction to radiation

A short film on background radiation from PHE can be found on You Tube[141]. A more detailed report is found on the government web site[142].

The NRPB booklet "Living with Radiation" (1998) is an excellent introduction to radiation dose and radiological protection. It is not available as a download but can still be bought.

The continued survival of many of the Nagasaki bomb survives is reported in a letter to the lancet[143].

A short review of radiation and its effects on life is given by the World Nuclear Organisation[144].

---

[137] http://www.atomica.co.uk/main.htm
[138] http://www.atomica.co.uk/shelters/main.htm
[139] https://www.ready.gov/nuclear-blast
[140] https://nepis.epa.gov/Exe/ZyPDF.cgi/P1004DE8.PDF?Dockey=P1004DE8.PDF
[141] https://www.youtube.com/watch?v=VCsJcfmL7TE&feature=player_embedded
[142] https://www.gov.uk/government/publications/ionising-radiation-exposure-of-the-uk-population-2010-review
[143] http://www.thelancet.com/pdfs/journals/lancet/PIIS0140-6736(15)00697-2.pdf
[144] http://www.world-nuclear.org/information-library/safety-and-security/radiation-and-health/radiation-and-life.aspx

E.3   Prior information

There are many examples of the public information letters sent to members of the public and businesses in DEPZ around the UK. These are just examples:

- BAE Systems Barrow in Furness[145],
- AWE[146],
- Sellafield[147],
- Portland[148],
- EDF, the public information for all EDF sites can be found on the EDF web-site[149].

E.4   Being prepared for emergencies

The Government generic advice on being prepared for an emergency is given on the Cabinet Office web-site[150].

Excellent and comprehensive advice from USA is available on the FEMA site[151].

More Concise and general advice for preparing for emergencies can be found in many other places on the internet. Examples I like are:

- Gloucestershire CC[152]
- Scottish Government[153]
- British Red Cross[154]
- Scottish Highlands[155]
- American web-site[156]

The government web-site for school emergency planning is at[157]

---

[145] http://www.cumbria.gov.uk/eLibrary/Content/Internet/536/5858/42053142012.pdf

[146] http://www.awe.co.uk/app/uploads/2014/07/REPPIR_May-2013.pdf

[147] http://www.sellafieldsites.com/wp-content/uploads/2012/07/What-to-do-in-an-Emergency-Booklet-Phase-1-Shelter-FINAL-061015.pdf

[148] https://www.dorsetforyou.gov.uk/media/163294/Information-in-the-event-of-a-Radiation-Emergency-in-Portland-Port/pdf/Radiation_booklet_04_02_2013_web_vers.pdf

[149] https://www.edfenergy.com/energy/safety-reporting/protecting-public

[150] https://www.gov.uk/government/publications/preparing-for-emergencies/preparing-for-emergencies

[151] https://www.fema.gov/media-library/assets/documents/7877#

[152] http://glosprepared.co.uk/be-prepared/preparing-your-family/

[153] http://www.readyscotland.org/at-home/

[154] http://www.redcross.org.uk/What-we-do/Preparing-for-disasters/How-to-prepare-for-emergencies

[155] http://www.redcross.org.uk/What-we-do/Preparing-for-disasters/How-to-prepare-for-emergencies

[156] https://www.ready.gov/prepare-for-emergencies

[157] https://www.gov.uk/guidance/emergencies-and-severe-weather-schools-and-early-years-settings

E.5 Countermeasures

The original NRPB Board Statement on Emergency Reference Levels (ERLs)[158]
Useful documents relating to shelter in place include:

- HM Government Shelter and Evacuation Guidance (rather long and technical)[159]
- The original NRPB document on Emergency Reference Levels[160]
- The Prior Information leaflets provided by site operators (see above)
- A Norfolk schools publication[161]
- A technical paper on the matter[162]

Information on the effectiveness of stable iodine can be found at:

- WHO[163]
- UK Working group (NRPB Document)[164]
- UK Iodine Dose Guidance[165]

Examples of the kinds of leaflets that would be given to people arriving at reception centres can be found on the internet:

- Dudley[166]
- Sussex[167]
- Gloucester[168]
- Japanese (within a guide to surviving earthquakes)[169]

Operational guidance for RMUs[170]

---

[158]
https://www.gov.uk/government/uploads/system/uploads/attachment_data/file/423879/NRPB_Vol_1_No_4_1990_-_for_website.pdf

[159]
https://www.gov.uk/government/uploads/system/uploads/attachment_data/file/274615/Evacuation_and_Shelter_Guidance_2014.pdf

[160]
https://www.gov.uk/government/uploads/system/uploads/attachment_data/file/423879/NRPB_Vol_1_No_4_1990_-_for_website.pdf

[161] http://www.schools.norfolk.gov.uk/view/NCC172136

[162] https://cfpub.epa.gov/si/si_public_file_download.cfm?p_download_id=478508

[163] http://www.who.int/ionizing_radiation/pub_meet/tech_briefings/potassium_iodide/en/

[164]
https://www.gov.uk/government/uploads/system/uploads/attachment_data/file/425072/Documents_of_the_NRPB_Volume_12_Number_3.pdf

[165] https://www.england.nhs.uk/wp-content/uploads/2015/04/potassium-iodate-85-adlts-child1.pdf

[166] http://www.dudley.gov.uk/resident/your-council/emergencies/emergency-plans/rest-centres/

[167] http://www.sussexemergency.info/glossary-and-faqs/what-can-i-expect-at-a-rest-centre-(rc)

[168]
http://www.gloucestershire.gov.uk/media/adobe_acrobat/0/6/Emergency%20Rest%20centre%20leaflet.pdf

[169]
http://www.metro.tokyo.jp/ENGLISH/GUIDE/BOSAI/FILES/01_Simulation_of_a_Major_Earthquake.pdf

An American version for comparison[171]

Information on protecting pets can be found:
- Greater Manchester[172]
- Somerset CC
  - Pets[173]
  - Ponies and horses[174]
  - Livestock[175]

E.6   Decontamination

Strategic national Guidance from 2004[176]
Advice from the British Red Cross[177]
Simple advice on self-decontamination[178,179]

E.7   Nuclear accidents

There is an awful lot of information about nuclear accidents available on the internet from a wide range of sources, some of which have scant regard for the facts. The sources I have used are:

- Windscale Fire Wikipedia[180]

- US-NRC TMI background[181]

- World Nuclear Organisation on TMI[182]

- FSA on Welsh Food Controls post Chernobyl[183]

170
https://www.gov.uk/government/uploads/system/uploads/attachment_data/file/340153/HPA-CRCE-017_for_website.pdf
171 https://emergency.cdc.gov/radiation/pdf/operating-public-shelters.pdf
172
http://www.gmemergencyplanning.org.uk/gmprepared/info/12/protecting_your_pet_during_an_emergency
173 http://www.somerset.gov.uk/EasySiteWeb/GatewayLink.aspx?alId=42846
174 http://www.somerset.gov.uk/EasySiteWeb/GatewayLink.aspx?alId=42845
175 http://www.somerset.gov.uk/EasySiteWeb/GatewayLink.aspx?alId=42850
176 https://www.gov.uk/government/uploads/system/uploads/attachment_data/file/62507/sng-decontamination-people-cbrn.pdf
177
http://www.redcross.org.uk/~/media/BritishRedCross/Documents/Archive/GeneralContent/C/Coping%20in%20a%20crisis.pdf
178 https://emergency.cdc.gov/radiation/selfdecon_wash.asp#how
179 https://emergency.cdc.gov/radiation/pdf/infographic_decontamination.pdf
180 https://en.wikipedia.org/wiki/Windscale_fire
181 http://www.nrc.gov/reading-rm/doc-collections/fact-sheets/3mile-isle.html
182 http://www.world-nuclear.org/information-library/safety-and-security/safety-of-plants/three-mile-island-accident.aspx

---

- WHO on Chernobyl[184]

- Parliamentary Brief on Chernobyl[185]

- World Nuclear Association on Fukushima[186]

The book *"Atomic Accidents, A History of Nuclear Meltdowns and Disasters from the Ozark Mountains to Fukushima"* by James Mahaffey analyses a surprisingly large numbers of accidents in a thought provoking way.

E.8   Nuclear emergency plans

The Office for Nuclear Regulation[187]
The National Risk Register of Civil Emergencies[188]
The UK's 36 nuclear licensed sites are listed at[189]
A guide to nuclear regulation in the UK[190]
The main regulations covering nuclear emergency plans are:
- Nuclear Industry Act (1965) As amended[191]
- The Radiation(Emergency Preparedness and Public Information) Regulations (2001)[192]

A rather dated overview of nuclear emergency planning can be found in HSE's 1994 publication "Arrangements for Nuclear Emergencies"[193]. A more up to date but rather government-centric document "concept of Operations"[194]
A brief review of the UK's nuclear emergency plans[195].

---

[183] https://www.food.gov.uk/science/research/radiologicalresearch/radiosurv/chernobyl
[184]

[185] http://researchbriefings.files.parliament.uk/documents/POST-PN-45/POST-PN-45.pdf
[186] http://www.world-nuclear.org/information-library/safety-and-security/safety-of-plants/fukushima-accident.aspx
[187] http://www.onr.org.uk/
[188]

https://www.gov.uk/government/uploads/system/uploads/attachment_data/file/419549/20150331_2015-NRR-WA_Final.pdf
[189] http://www.onr.org.uk/licensees/pubregister.pdf
[190] http://www.onr.org.uk/documents/a-guide-to-nuclear-regulation-in-the-uk.pdf
[191] http://www.legislation.gov.uk/ukpga/1965/57
[192] http://www.hse.gov.uk/radiation/ionising/reppir.htm
[193] http://www.onr.org.uk/arrangement-for-nuclear-emergencies.pdf
[194]

https://www.gov.uk/government/uploads/system/uploads/attachment_data/file/472419/NEPRG00_-_Concept_of_Operations.pdf
[195]

https://www.gov.uk/government/uploads/system/uploads/attachment_data/file/467198/Nuclear_emergencies_-_information_for_the_public_October_2015.pdf

---

Current BEIS Nuclear emergency planning guidance to help local planners, Whitehall departments, devolved administrations and agencies write effective plans can be found on their website[196].

Many on-site plans (public versions) can be found on the internet. For example:
- Dungeness B[197]
- Dounreay[198]

Public versions of many of the off-site plans can be found on the internet. For example: Aldermaston[199], Highland[200], Sizewell[201], Hunterston[202].

Resources for community resilience and emergency planning[203].

E.9   Radiological protection

A brief review of the discovery and early use of x-rays can be found on the British Library web-site[204].

A good review of the early history of radiological protection[205].

Since 1928, ICRP has developed, maintained, and elaborated the International System of Radiological Protection used world-wide as the common basis for radiological protection standards, legislation, guidelines, programmes, and practice. Their web site[206] links to a wealth of information but most of it is very technical. In the UK the profession is supported by the Society for Radiological Protection (SRP)[207].

The main regulations for radiological protection are:

- Ionising Radiation Regulations 1999[208]
- Management of Health and Safety at Work Regulations 1999[209]

---

[196] https://www.gov.uk/government/publications/national-nuclear-emergency-planning-and-response-guidance

[197] https://www.edfenergy.com/file/1475/download

[198] http://www.dounreay.com/UserFiles/File/Emergency%20arrangements/Dounreay%20Emergency%20Plan%20Issue%2010(2).pdf

[199] http://info.westberks.gov.uk/CHttpHandler.ashx?id=42490&p=0

[200] http://www.highland.gov.uk/info/1226/emergencies/72/emergency_planning

[201] http://www.suffolkresilience.com/emergency-plans/sizewell-offsite-plan/

[202] https://www.east-ayrshire.gov.uk/Resources/PDF/H/Hunterston-off-site-emergency-plan-redacted-version.pdf

[203] https://www.gov.uk/government/publications/community-resilience-resources-and-tools

[204] http://www.bl.uk/learning/cult/bodies/xray/gallery/xraygallery.html

[205] http://www.physics.isu.edu/radinf/50yrs.htm

[206] http://www.icrp.org/

[207] https://srp-uk.org/

[208] http://www.legislation.hmso.gov.uk/si/si1999/19993232.htm

- The Justification of Practices involving Ionising Radiations Regulations 2004[210]

The UK Strategy for radioactive discharges (2009)[211].

Detailed information about annual radioactive discharges, the measurements of contaminants in the environment and the estimation of doses to the critical group can be found in the annual RIFE Reports (Radioactivity in Food and the Environment)[212].

[209] http://www.legislation.hmso.gov.uk/si/si1999/19993242.htm
[210] http://www.opsi.gov.uk/cgi-bin/htm_hl.pl?DB=opsi&STEMMER=en&WORDS=justif+practic+&COLOUR=Red&STYLE=s&URL=http://www.opsi.gov.uk/si/si2004/20041769.htm
[211] https://www.gov.uk/government/uploads/system/uploads/attachment_data/file/249884/uk_strategy_for_radioactive_discharges.pdf
[212] https://www.food.gov.uk/science/research/radiologicalresearch/radiosurv/rife

E.10   Lexicon

The nuclear industry and radiological protection use a lot of technical terms. Some are listed and explained below. Many of these definitions can be traced to the national lexicon of UK civil protection terminology[213].

Term	Definition
Absorbed dose (Gy)	The radiation energy imparted per unit mass of an irradiated body. It is measured in joule per kilogram, a unit which is also called the gray (Gy)
Alpha radiation	A heavy particulate radiation (a helium nucleus) emitted by a number of heavy elements it has very short range but can cause considerable damage if emitted in or very near living matter due to localized energy deposition.
Averted dose	The dose saved by implementing a countermeasure.
Beta radiation	A light particulate radiation (an electron) emitted by neutron heavy nuclei. It has a short range but can penetrate the outer layers of skin and damage the underlying living tissue.
Casualty Bureau	Initial point of contact for receiving and assessing information about casualties.
Containment vessel	A reinforced structure built around a nuclear reactor with the design intent of retaining the majority of radioactivity released from the reactor in an accident.
Contamination	In this sense, the presence of radioactive material, usually in the form of dust, on surfaces or people and/or their clothes.

213

https://www.gov.uk/government/uploads/system/uploads/attachment_data/file/128797/LEXIC
ON_v2_1_1-Feb-2013.xls

Term	Definition
Decay heat	The name given to the heat generated by the decay of fission products in the fuel assembly. It is because of this that reactor cores must continue to be cooled after shutdown.
Decontamination	Removal or reduction of hazardous materials to lower the risk of further harm to victims and/or cross contamination.
Detailed Emergency Planning Zone	An area around a nuclear licensed site for which a detailed off-site (local authority) plan should exist.
Deterministic Effect	An effect of radiation which has a threshold dose below which the effect will not be seen but which then gets more severe as the dose increases. Sun burn is an example of a deterministic effect. (See also Stochastic Effect)
Effective dose	The effective dose is defined by the ICRP as the sum of the equivalent doses in the principal tissues and organs in the body, each weighted by a tissue weighting factor, w_T. Measured in Sieverts (Sv). It is a measure of the harm the radiation does to the person.
Emergency exposure situation (ICRP 103)	Unexpected situations that may require urgent protective actions and perhaps also longer-term protective actions to be considered.
Equivalent dose	Multiplying the absorbed dose by appropriate weighting factors depending on the type of radiation, creates the equivalent dose in the relevant organ or tissue. Measured in Sieverts (Sv). It is a measure of the harm the radiation does to the organ or tissue.
Evacuation	Removal, from a place of actual or potential danger to a place of relative safety, of people and (where appropriate) other living creatures.

Term	Definition
Fallout	A term applied to radioactive dust settling on the ground and other surfaces following a release of radioactivity - usually applied to the aftermath of a nuclear bomb.
Family and Friends Reception Centre	A facility to help reunite family and friends with survivors.
FEMA	Federal Emergency Management Agency (USA). Mission: *To lead America to prepare for, prevent, respond to and recover from disasters with a vision of "A Nation Prepared"*.
Food Controls	The combination of legal controls (under FEPA) and advice not to eat food that might have been contaminated by a release of radioactivity.
Gamma radiation	A photon (light energy) emitted by many radioactive materials when they decay. Very long range and deposit energy in a spread-out manner.
Go in, Stay in, Tune in	Generic advice to members of the public on how to respond to an emergency. The intention is to get them into buildings that protect them from the hazard and provide them with information on broadcast media.
Half-life	The time taken for half of the radioactive atoms present to decay. This varies for different isotopes from a very small fraction of a second to many millions of years. The fraction left after each successive half-life is ½, ¼, 1/8, 1/16, 1/32 .....
IAEA	International Atomic Energy Agency
Ionising radiation	Types of radiation (including alpha, beta and gamma) that can ionize the material they pass through.
INES	International Nuclear Event Scale
LBR	Lifetime baseline risk

Term	Definition
Medical exposure	Exposure to ionizing radiations of persons as part of their diagnosis or treatment. (ICRP)
Moderator	A material in a nuclear reactor that is there to reduce the speed of neutrons produced in fission to make it more likely that they will induce further fissions. Hydrogen, in the form of water is a common moderator. Graphite (carbon) is used in the British Gas Cooled Reactors.
Occupational exposure	Exposure to radiation incurred at work, and principally as a result of work. (ICRP)
ONR	Office for Nuclear Regulation - whose mission is: "to provide efficient and effective regulation of the nuclear industry, holding it to account on behalf of the public".
Projected dose	The overall exposure to a member of the public in an accident should no protective actions be taken. See residual dose and averted dose
Public Exposure	Exposure to ionizing radiation to members of the public in their everyday life (not at work or undergoing medical exposure). (ICRP)
Radioactive Dispersal Device (RDD)	Dirty Bomb - device intended to disperse a radioactive source in public area.
Radioactive decay	The process whereby an unstable atomic nucleus changes its form and loses energy in the form of radiation.
Radioactivity	The particles or photons (light) which are emitted when unstable atoms decay.
Reasonably foreseeable	According to REPPIR guidance (paragraph 50) "In the context of a radiation emergency, a reasonably foreseeable event would be one which was less than likely but realistically possible".

Term	Definition
REPPIR	The Radiation (Emergency Preparedness and Public Information) Regulations, 2001.
Rest Centre	A facility managed by the local authority for temporary accommodation of evacuees and homeless survivors. With overnight facilities.
Residual dose	The dose to which a member of the public would be exposed to in an accident if countermeasures were taken to reduce the dose. See projected dose and averted dose
RMU	Radiation Monitoring Unit
Shelter	Countermeasure advice for the public to 'Go in, stay in and tune in'. Includes closing windows and doors to minimise the ingress of hazardous materials.
Stable Iodine (Potassium Iodate) tablets	A stable form of iodine, which can bind to and protect the thyroid gland, preventing uptake of radioactive iodine by the gland and thus reduce the likelihood of thyroid cancer in the future.
Stochastic Effect	A radiation effect for which the probability of the effect being seen is proportional to the dose and for which the dose has no impact on the severity of the effect. Cancer is a stochastic effect. (See also Deterministic Effect)
Survivor Reception Centre	A secure area to which survivors not requiring acute hospital treatment can be taken for short-term shelter and first aid.
WHO	World Health Organisation - whose "primary role is to direct and coordinate international health within the United Nations' system".

Appendix F - Your Home Resilience Plan

This section aims to help you to think about and complete your own home resilience plan.

F.1   Risk Assessment

The first step is to ensure that you make the right plans for your situation. There is no need to have a hurricane plan if you live in Gloucestershire in the UK. So consider what the risks you face in everyday life given where you live and work and what you do? Consider (for example):

- Floods damaging home;
- Floods cutting off roads and services;
- Power cut;
- Failure of mobile phone (Cell phone) system;
- Transport disruption (strikes, bad weather, accident);
- Nuclear accident (do you live or work near a nuclear site?);
- Extreme cold or hot weather;
- Severe storm;
- House/place of work/school on fire.

Consider how likely the hazard or emergency is to affect you and if it is high then consider what your first actions might be if given a warning or if the event strikes without warning. One method for doing this (see Table 5) is to consider how often an event might happen and give it a probability score. For example something that might be expected to happen no more than once in 50 years could score 1, something that might happen about once every 10 years scores 2 and something that might be expected to happen once a year or more scores 3. Then consider how disruptive the event would be if you were unprepared. So a few hours disruption scores 1, a day of disruption scores 2 and anything more scores 3.

By multiplying the two scores (frequency and disruption) you have a single number - often called the "risk". The hazards or emergencies with the highest scores or the ones to play most attention to.

Now consider if there is anything you can do to reduce the probability or reduce the disruption. If power cuts are frequent could you install a generator or batteries to reduce the probability or stock up on camping lights, heaters, cookers and food to reduce the disruption?

Utilities - Do you know how to turn off your home's electricity, gas and/or water supplies should that be needed?

You may have to turn off the gas if you smell gas and the water if you have a bad non-isolatable leak.

Do you know the basics of your children's school's emergency plan and your role in them?

If you take any such steps you can reduce the scores for that hazard or emergency and determine the highest risks remaining.

The winter of 2015/2016 was the second wettest winter on record in the UK and a series of storms resulted in heavy and sustained rainfall. 17,600 UK properties were flooded and several bridges collapsed, disrupting access to and from local communities. 61,000 people lost power due to flooding in the Lancaster area. Economic damage was estimated to be about £1.6 billion.

F.2   Flood planning

For floods you should look at the risks of your area flooding and the nature of the possible floods. In the UK you can look up your area's flood risk on the gov.uk website[214]. For USA flood risk products see the FEMA web-site[215] . Are you likely to be cut off because roads around you flood? Is your house liable to flood? If the answer to either of these questions is "yes" then it would be sensible to plan accordingly.

A simple flood plan is outlined on the UK government website[216]. You may wish to consider what you could do to protect your home and belongings from flood before any alert - How best to stop water entering your property, where to get sandbags or other water barriers if prone to flooding, who could help you, what might you move before evacuating and what might you take with you. Have a plan for what you might do if given a few hours' notice of a rising flood. Do you know how to switch off the electricity?

A great deal of advice is available to those caught in floods in the UK on the Government website[217]. An important message is that flood water is often not at all clean and contact should be avoided.

---

[214] https://flood-warning-information.service.gov.uk/long-term-flood-risk/map

[215] https://www.fema.gov/risk-map-flood-risk-products

[216]

https://www.gov.uk/government/uploads/system/uploads/attachment_data/file/444659/LIT_41 12.pdf

[217]   https://www.gov.uk/government/collections/flooding-health-guidance-and-advice#people-living-through-a-flood

F.3   Utilities

Can you cope in a power cut? Do you have battery or wind-up lights and radio?
Do you have a landline phone that will work without mains electricity or wind up
chargers for your mobile phones? How will you keep warm and fed without
electricity and/or gas? Remember that if the power cut affects your local
supermarket they may be unable to sell anything due to the lack of tills.

	What to do (Phone numbers for UK)	Contact number (UK)
Gas	If you smell gas[218], call free on 0800 111 999. • Turn off the gas at the meter, unless the meter is located in a cellar or basement - in which case, do not enter. • If there is a smell of gas in the cellar or basement, you should evacuate the building. • Extinguish all naked flames - do not smoke or strike matches. • Turn off all gas appliances and do not use until they are checked by the engineer. • Do not operate any electrical appliances or turn any switches on or off. • Open doors and windows to ventilate the property. • Keep people away from the area. • Immediate access will be required	UK Gas emergency number 0800 111 999
Electricity	Power cut • Switch off anything electrical that should be left unattended (rings on cooker, for example) • Leave a light on to show you when power is back on. • Check to see if your neighbours are okay. • Wrap up warm (if needed).	UK: Contact National Grid on 105 to report cut (free of charge)
Water	If your house is flooding turn the electricity off at the mains.	
	For water supply or wastewater emergencies, blocked drains or possible pollution, call: 0330 303 0368 (Calls charged at local rate)	

---

[218] http://www.wwutilities.co.uk/services/smell-gas/

F.4   Drinking water interruption

There are various estimates of the amount of water a person needs, either to just survive or to be reasonably comfortable. Needs include drinking water, water to cook with and water to keep self and environment clean. A US EPA document[219] puts the needs in the range of half a gallon to 5 gallons a day (2 - 20 litres). In the UK the Pitt Report[220] following the floods in England in 2007 recommended a minimum of 20 litres per day up from the 10 litres requirement that had been in place. In these floods some people were without mains water for 17 days.

In both the USA and UK there are regulations and guidance requiring plans to be in place to provide water to populations for whom the normal supply is unavailable or unusable. These plans are based on risk assessments which consider why water may not be available (earthquakes and hurricanes feature west of the Atlantic while floods and technical failures head the risks in the UK).

In the UK the public water companies are required to plan to supply each customer with 10 litres of drinkable water per person per day for the first 24 hours and 20 litres per day thereafter. This is usually delivered by tanker, in which case you take your own container to collect your water, or by bottle, which may be delivered to your door or to a neighbourhood distribution point. This water must be "wholesome at the point of supply" although advice may be given to boil the water before drinking it[221].

The UK generic guidance on Planning for Major Water and Wastewater Incidents in England and Wales can be found on the .gov.uk website[222].

Even though there are credible plans to provide you with an amount of water in an emergency, there may be a delay before this bottled or bowsered water is available. Having some bottled water in the house to cover this period is sensible. (The US FEMA recommend one gallon (4 litres) per person per day for three days[223]) If you do keep a stock of bottled water remember to store it carefully and to replace it at suitable intervals.

---

[219]

https://www.awwa.org/Portals/0/files/resources/water%20knowledge/rc%20emergency%20prep/Emergencywater.PDF

[220]

http://webarchive.nationalarchives.gov.uk/20100812084907/http:/archive.cabinetoffice.gov.uk/pittreview/_/media/assets/www.cabinetoffice.gov.uk/flooding_review/pitt_review_full%20pdf.pdf

[221]http://dwi.defra.gov.uk/stakeholders/guidance-and-codes-of-practice/SEMD%20DIRECTION%201998.pdf

[222]

https://www.gov.uk/government/uploads/system/uploads/attachment_data/file/62044/water_guidance.pdf

[223] https://www.ready.gov/build-a-kit

If storing water in your own bottles for any length of time then: wash bottles thoroughly in hot water. Fill each bottle with tap water until it overflows. Add five drops of household bleach per litre of water (or half a teaspoon for 10 litres) and Store in a cool dark place and replace the water every 12 months. It is probably easier to buy bottled water and use and replace every few months.

F.5　Lost phone

Although many people carry mobile phones with their friends' and family's contact details in them - could you cope if the mobile phone network was down or if you'd lost your phone? (How many important phone numbers can you actually remember?). Having a written contact list at home and a more focused one in your pocket or bag could be useful.

For transport disruptions there is the initial concern of getting everyone home or to a place where they can stay. A contact list is helpful in these circumstances.

For alerts of bad weather over a period of days you would be advised to take steps to minimise your travel needs. Stock up on food and water and check your power cut planning.

F.6　Local radio station

The local radio station would be covering any major local new story and are a good source of information. Do you know their frequency? Have you got a battery or wind up radio to use if the power is cut?

F.7　Family safe place / meeting place

If the family home is not available for any reason then where are the family going to meet and where are they going to stay? Do you have family or friends living far enough from you not to be affected by the same event but near enough for you all to reach? Is there a coffee shop, or other warm place, that you could agree to wait for each other in?

If you have to evacuate your home do you know how to do it? (What to take with you, where the things you need are kept, how to leave the home (lights and heating off, windows closed, doors locked, fridges and freezers on unless there is a risk of flooding), how to travel to and the best routes out under different circumstances).

F.8   Family documents

Documents - Where are your insurance policies, birth certificates, marriage certificates, passports, savings books etc.? Can you find them if you have to leave quickly? Are they safe from water damage (floods or firefighting)? It is a good idea to keep these key documents together in a water tight folder ready to be collected in an emergency.

Figure 10 Key documents in waterproof folder

F.9   Special needs

Do any family members have special needs? Can they cope in an emergency?

- Mobility impaired, less able to make home safe in times of need or unable to evacuate;
- Hearing impaired, unable to hear normal warning systems;
- Sight impaired, less able to manage alone;
- Home dialysis patient, more dependent on water and electricity supplies;
- Supported in home such as regular user of services to provide meals, hygiene and living support?
- Non-English speaking persons who may need help understanding situation and appropriate response;

- People without vehicles, may need more help in an evacuation;
- People with special dietary needs or dependent on medicines who may have supply issues.

Consider if your family has any special needs and work out in advance how best to support them in a crisis. UK utilities have registration systems for people with such needs to make themselves known so that they get the extra help they need in times of crisis. Consider, for example, signing onto the electricity users Priority Services Register[224] (UK).

F.10 Home Emergency Kit

It is useful to have an emergency kit prepared for reasonably foreseeable emergencies this could include:
- Water for drinking and sanitation (4 litres per day per person);
- Food, three days supply of non-perishable food (tinned food or dried food for example - make sure you have a tin opener if needed!);
- A battery powered or wind-up radio;
- Torch and batteries, or wind up torch and lights;
- First Aid kit;
- Prescription medicines;
- Key family documents;
- Credit cards and cash;
- Sleeping bags and spare clothes (suitable for climate);
- Sanitary equipment and supplies (soap, toothpaste & brushes, toilet paper, feminine and personal hygiene items etc.);
- Plastic bags for wastes;
- Iodine tables (if issued).

Different contents are required for kits to allow you to shelter at home or to evacuate.

---

[224] https://www.ofgem.gov.uk/consumers/household-gas-and-electricity-guide/extra-help-energy-services/priority-services-register-people-need.

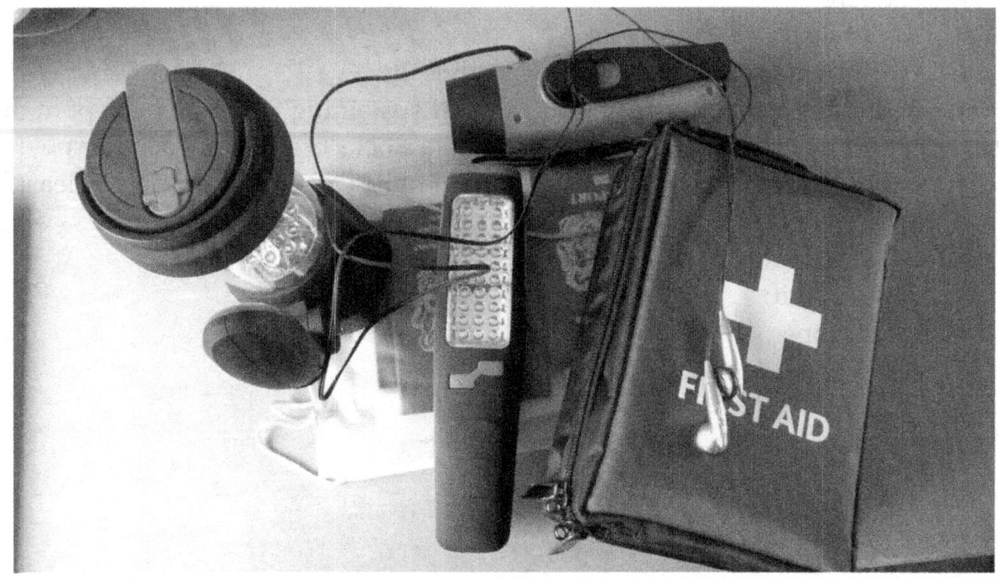

Figure 11 - Emergency Kit (Part of)

Wind up radio and torch, wind up and battery torches, first aid kit, document wallet.

F.11   Car Emergency Kit

When travelling in the car a breakdown or traffic jam can easily add several hours to the journey time, sometimes more. Depending on the weather (too hot or too cold) it can quickly become uncomfortable for those not prepared.

Getting trapped in snow or a flood in an isolated area can quickly become very uncomfortable and dangerous. It is better to plan or delay your journey to prevent this from happening.

The probability of a breakdown can be reduced by servicing your vehicle.

Depending on the season the following might be part of your car emergency kit:
- A blanket each for the number of people who often use the car (you might have to stay in the car for a few hours in traffic jams or breakdown situations) + warm clothes. (If you breakdown on a motorway you are advised to leave your car and stand away from the road. It could be a while);
- High visibility vests for the number of people who often use the car;

- Rain poncho(s);
- A torch;
- A mobile phone;
- Warning triangles - at least two is advised;
- Some chocolate bars, fruit bars or other sustaining food that be kept in the car for a few months or ensure you take stocks of food for each journey;
- Enough drinking water for everybody in the car - enough to last at least few hours;
- Snow shovel;
- First Aid kit;
- Ice scrapper;
- A paper road atlas;

F.12 Risk Assessment forms and family emergency plan templates

Risk Assessment

	Probability	Consequence
1	Once every 50 years	Minor disruption
2	Once every 10 years	Need to take action (few days to fix)
3	About once a year	Several weeks or more disruption

Hazard or emergency	Before risk reduction			Risk Reduction (what can you do now to reduce the risk?)	After risk reduction		
	Prob (P)	Cons (C)	Risk (PxC)		Prob (P)	Cons (C)	Risk (PxC)

Table 5 Home risk assessment

Contact list - Include all those who live with you and family and friends.

Name	Mobile number	Landline number	Work place/ school

Local Radio Station Frequency

<table>
<tr><td>

</td></tr>
</table>

General contacts	Company Name	Telephone
Floodline		
Electricity provider		
Gas provider		
Water company		
Telephone provider		
Insurance company (note policy number)		

Utilities

	How to switch off if needed
Electricity	
Gas	
Water	

Family place to meet if house is unreachable (friend or relative's house, coffee shop etc.) Include telephone number so that you can leave messages there before arriving.

<table>
<tr><td>Near to home

</td></tr>
<tr><td>Outside immediate area

</td></tr>
<tr><td>

</td></tr>
</table>

Location of stable iodine tablets (if you have been issued with them) and emergency kit.

<table>
<tr><td>

</td></tr>
</table>

Family's key information

Name	Date of Birth	National Insurance Number	Passport number

Friends or neighbours who may need extra help

Name	Address	Phone

Notes:

Also By Keith Pearce
Nuclear Emergency Planning for Local Authorities.

What is it about?
This book provides an introduction to nuclear physics, reactor physics, radiation protection and the UK emergency response organisation.

Who should read it This book should be read by people who, although not nuclear industry specialists, may have to respond to an off-site nuclear emergency in some capacity.

Available from Amazon: https://amzn.to/2GVi38C

How to Survive a Nuclear Emergency: Published by Katwab Limited. Registered in England and Wales. Company Number 10545578
ISBN 9781976786938

Made in the USA
Monee, IL
07 July 2026

56544272R00063